Water Harvesting

Ecological and Economic Appraisal

Water Harvesting

Ecological and Economic Appraisal

Archana Mishra

AUTHORSPRESS
Publishers of Creative & Scholarly Books

Worldwide Circulation through Authorspress Global Network
First Published in 2006
by
Authorspress
Q-2A Hauz Khas Enclave, New Delhi-110 016 (India)
Phone: (0) 9818049852
E-mail: authorspressgroup@gmail.com
Website: www.authorspressbooks.com

Water Harvesting:
Ecological and Economic Appraisal

ISBN 81-7273-307-0

Printed in India at Kensington Press

Preface

Water harvesting is the activity of direct collection of rainwater that would otherwise have gone down the drainage system, into the ground or been lost to the atmosphere through evaporation. As an effective method of utilising rainwater for domestic and agricultural use, it is already widely used throughout the world. It is a method which has been used since ancient times and is increasingly being accepted as a practical method of providing potable water in development programmes throughout the world. It has wide application also in urban and peri-urban areas where the reliability and quality of piped water is increasingly being questioned. Despite these facts the percentage of households using rainwater harvesting both in urban and rural areas is remarkably low.

Large numbers of population are living in the arid and semi-arid belts of the world where rainfall is limited and very irregular. During good years, agriculture in these areas produces satisfactory yields. However, reduced rainfall or prolonged dry spells are leading immediately to crop failure. Irrigation would be the logic solution but water scarcity and the high cost of irrigation infrastructure are major handicaps. Water harvesting is a proven technology to increase food security in drought prone areas. Erosion control and recharge of ground water are additional advantages of water harvesting techniques. Water harvesting means to understand the value of rain, and to make optimum use of the rainwater at the place where it falls. It promotes self-sufficiency and fosters an appreciation for water as a resource. It also conserves energy as the energy input needed to operate a centralised water system is bypassed.

Today, countries across the world promote water harvesting as the principal means of supplying household water. However in most countries, extension and irrigation staff and general public have often very limited knowledge about the various water harvesting techniques and the associated socioeconomic implications. The dissemination and adoption of the technology on a larger scale by farmers is hampered by this lack of trained field staff.

This book has been written with the intention of providing technicians, extension workers, rural development planners, farmers and general public with practical guidelines for developing and implementing of water harvesting programmes. It presents the scientific basics, perspective and technical information of traditional and modern water harvesting practices. It includes a good number of tables, diagrams, case studies of water harvesting practices and detailed technical designs for various systems. The ecological and economic benefits of water harvesting techniques are also highlighted. It is hoped that the book will serve as a useful guide for proper development and implementation of water harvesting programmes.

Archana Mishra

Contents

CHAPTER 1

Introduction to Water Harvesting

Water harvesting is the collection of runoff for productive purposes. Instead of runoff being left to cause erosion, it is harvested and utilized. In the semi-arid drought-prone areas where it is already practised, water harvesting is a directly productive form of soil and water conservation. Both yields and reliability of production can be significantly improved with this method. Water Harvesting (WH) can be considered as a rudimentary form of irrigation. The difference is that with WH the farmer has no control over timing. Runoff can only be harvested when it rains.

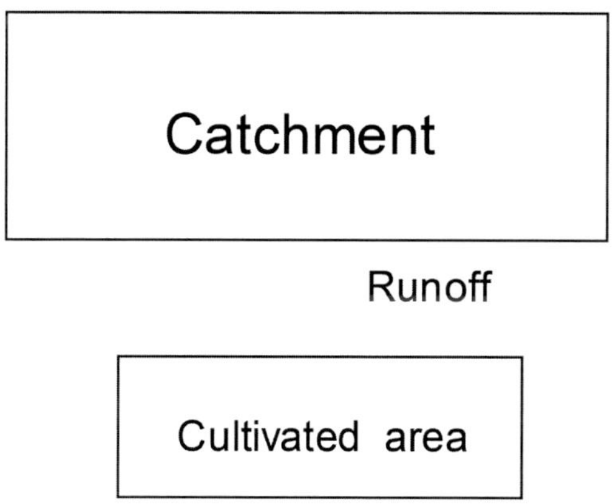

Figure 1. The principle of water harvesting

Various forms of water harvesting (WH) have been used traditionally throughout the centuries. Some of the very earliest agriculture, in the Middle East, was based on techniques such as diversion of "wadi" flow onto agricultural fields. In the Negev Desert of Israel, WH systems dating back 4000 years or more have been discovered. These schemes involved the clearing of hillsides from vegetation to increase runoff, which was then directed to fields on the plains.

Floodwater farming has been practised in the desert areas of Arizona and northwest New Mexico for at least the last 1000 years. The Hopi Indians on the Colorado Plateau, cultivate fields situated at the mouth of ephemeral streams. Where the streams fan out, these fields are called "Akchin". Pacey and Cullis describe microcatchment techniques for tree growing, used in southern Tunisia, which were discovered in the nineteenth century by travellers.

In the "Khadin" system of India, floodwater is impounded behind earth bunds, and crops then planted into the residual moisture when the water infiltrates. The importance of traditional, small scale systems of WH in Sub-Saharan Africa is just beginning to be recognized. Simple stone lines are used, for example, in some West African countries, notably Burkina Faso, and earth bunding systems are found in Eastern Sudan and the Central Rangelands of Somalia.

Water harvesting technology is especially relevant to the semi-arid and arid areas where the problems of environmental degradation, drought and population pressures are most evident. It is an important component of the package of remedies for these problem zones, and there is no doubt that implementation of WH techniques will expand.

Water Harvesting Techniques

Traditional water management techniques most of them being simple, sure to implement and of low capital investment. The classical sources of irrigation water are often at the break of overuse and therefore untapped sources of (irrigation) water have to be sought for increasing agricultural productivity and providing sustained economic base. Water harvesting for dry-land agriculture is a traditional water management technology to ease future water scarcity in many arid and semi-arid regions of world. This old technology is gaining new popularity these days. As the appropriate choice of technique depends on the amount of rainfall and its distribution, land topography, soil type and soil depth and local socio-economic factors, these systems tend to be very site specific.

The water harvesting methods applied strongly depend on local conditions and include such widely differing practices as bunding, pitting, microcatchments water harvesting, flood water and ground water harvesting. There are three types of water harvesting techniques

Rainwater Harvesting

Rainwater harvesting is defined as a method for inducing, collecting, storing and conserving local surface runoff for agriculture in arid and semi-arid regions. Three types of water harvesting are covered by rainwater harvesting. a) Water collected from roof tops, courtyards and similar compacted or treated surfaces is used for domestic purpose or garden crops. b) Micro-catchment water harvesting is a method of collecting surface runoff from a small catchment area and storing it in the root zone of an adjacent infiltration basin. The basin is planted with a tree, a bush or with annual crops. c) Macro-catchment water harvesting, also called harvesting from external catchments, is the case where runoff from hill-slope catchments is conveyed to the cropping area located at hill foot on flat terrain.

Flood Water Harvesting

Flood water harvesting can be defined as the collection and storage of creek flow for irrigation use. Flood water harvesting, also known as 'large catchment water harvesting' or 'Spate Irrigation', may be classified into following two forms: a) In case of 'floodwater harvesting within stream bed', the water flow is dammed and as a result, inundates the valley bottom of the flood plain. The water is forced to infiltrate and the wetted area can be used for agriculture or pasture improvement. b) In case of 'floodwater diversion', the wadi water is forced to leave its natural course and conveyed to nearby cropping fields.

Groundwater Harvesting

Groundwater harvesting is a rather new term and employed to cover traditional as well as unconventional ways of ground water extraction. Qanat systems, underground dams and special types of wells are few examples of the groundwater harvesting techniques. For example, Qanats, widely used in Iran, Pakistan, North Africa and even in Spain, consists of a horizontal tunnel that taps underground water in an alluvial fan, brings it to the surface due to gravitational effect. Qanat tunnels have an inclination of 1-2% and a length of up to 30 km. Many are still maintained and deliver steadily water to fields for agriculture production and villages for drinking water supply.

Groundwater dams like 'Subsurface Dams' and 'Sand Storage Dams' are other fine examples of groundwater harvesting. They obstruct the flow of ephemeral streams in a river bed; the water is stored in the sediment below ground surface and can be used for aquifer recharge. Sand filled reservoirs have the following advantages:

i) Evaporation losses are reduced,
ii) no reduction in storage volume due to siltation,
iii) stored water is less susceptible to pollution, and
iv) health hazards due to mosquito breeding are avoided.

In most rain and floodwater harvesting schemes the water delivered by surface runoff and overland flow is stored (only) in the soil matrix. This means, that its application is limited to the rainy season. To allow cropping outside the rainy season, a number of storage media are employed, ranging from ferrocement tanks of a few m^2 content to large reservoirs, storing millions of m^3. In India and Sri Lanka, more than 500000 tanks store rain water, sometimes supplemented by water from streams or small rivers.

Tanks play several important roles e.g. as flood-control system and in preventing soil erosion and wastage of runoff during periods of heavy rainfall. Additionally, they recharge the groundwater in surrounding areas. The larger ones, 10 to 30 hectares in size, feed several thousand hectares of irrigated land. They are equipped with sluices, which deliver water to an extensive canal system. Without this tank system, paddy cultivation in large parts of the country would be impossible. These rainwater reservoirs are not only employed for irrigation in arid or semi-arid regions, but in semi-humid areas (up to 1300 mm/a rainfall), too.

As several disadvantages are connected with surface storage of water - large evaporation losses, loss of storage caused by siltation, pollution problems and loss of agricultural land, underground storage of water may be an interesting alternative. This storage can be done in near surface aquifers (e.g. in wadi beds), calling for a conjunctive management of water resources, or in cisterns. Cisterns are man-made caves or underground constructions to store water. Often the walls of these cistern are plastered; their water losses by deep percolation or by evaporation can be minimal. The construction of cisterns was already practised several thousand years ago; chalky rocks were preferred. Traditionally, in Mediterranean houses, one cellar room was specifically designed to store rainwater. Similar in-house cisterns are known from Rajasthan, NW India. In the same region, 'Kunds', covered underground tanks with a plastered catchment, are found. Nowadays cisterns are often constructed using concrete.

Identification of Rain and Flood Harvesting Areas

The most important parameters to be considered in identifying areas suitable for rain and floodwater harvesting are as follows:

a) *Rainfall*: The knowledge of rainfall characteristics (intensity and distribution) for a given area is one of the pre-requisites for designing a water harvesting system. The availability of rainfall data series in space and time and rainfall distribution are important for rainfall-runoff process and also for determination of available soil moisture. A threshold rainfall events (e.g. of 5 mm/event) is used in many rainfall runoff models as a start value for runoff to occur. The intensity of rainfall is a good indicator of which rainfall is likely to produce runoff, Useful rainfall factors for the design of a rain- or floodwater harvesting system include:

   i) Number of days in which the rain exceeds the threshold rainfall of the catchment, on a weekly or monthly basis.

   ii) Probability and occurrence for the mean monthly rainfall.

   iii) Probability and reoccurrence for the minimum and maximum monthly rainfall.

   iv) Frequency distribution of storms of different specific intensities.

b) *Land use or vegetation cover*: Vegetation is another important parameter that affects the surface runoff. In West Africa and Syria proved that an increase in the vegetation density results in a corresponding increase in interception losses, retention and infiltration rates which consequently decrease the volume of runoff. Vegetation density can be characterised by the size of the area covered under vegetation. There is a high degree of congruence between density of vegetation and suitability of the soil to be used for cropping.

c) *Topography and terrain profile*: The land form along with slope gradient and relief intensity are other parameters to determine the type of water harvesting. The terrain analysis can be used for determination of the length of slope, a parameter regarded of very high importance for the suitability of an area for macro-catchment water harvesting. With a given inclination, the runoff volume increases with the length of slope. The slope length can be used to determine the suitability for macro or micro- or mixed water harvesting systems decision making.

d) *Soil type & soil depth*: The suitability of a certain area either as catchment or as cropping area in water harvesting depend strongly on its soils characteristics viz.

   i) surface structure; which influence the rainfall-runoff process,

   ii) the infiltration and percolation rate; which determine water movement into the soil and within the soil matrix, and

   iii) the soil depth incl. soil texture; which determines the quantity of water which can be stored in the soil.

e) *Hydrology and water resources*: The hydrological processes relevant to water harvesting practices are those involved in the production, flow and storage of runoff from rainfall within a particular project area. The rain falling on a particular catchment area can be effective (as direct runoff) or ineffective (as evaporation, deep percolation). The quantity of rainfall which produces runoff is a good indicator of the suitability of the area for water harvesting.

f) *Socio-economic and infrastructure conditions*: The socio-economic conditions of a region being considered for any water harvesting scheme are very important for planning, designing and implementation. The chances for success are much greater if resource users and community groups are involved from early planning stage onwards. The farming systems of the community, the financial capabilities of the average farmer, the cultural behaviour together with religious belief of the people, attitude of farmers towards the introduction of new farming methods, the farmers knowledge about irrigated agriculture, land tenure and property rights and the role of women and minorities in the communities are crucial issues.

g) *Environmental and ecological impacts*: Dry area ecosystems are generally fragile and have a limited capacity to adjust to change. If the use of natural resources (land and water), is suddenly changed by water harvesting, the environmental consequences are often far greater than foreseen. Consideration should be given to the possible effect on natural wetlands as on other water users, both in terms of water quality and quantity.

New water harvesting systems may intercept runoff at the upstream part of the catchment, thus depriving potential down stream users of their share of the resources. Water harvesting technology should be seen as one component of a regional water management improvement project. Components of such integrated plans should be the improvement of agronomic practices, including the use of good plant material, plant protection measures and soil fertility management.

Water Harvesting Systems for Plant Production

There are three basic categories of waterharvesting systems for plant production:

Microcatchments

Sometimes it also referred to as "Within-Field Catchment System". The following are the main characteristics:

i) overland flow harvested from short catchment length
ii) catchment length usually between 1 and 30 metres
iii) runoff stored in soil profile
iv) ratio catchment: cultivated area usually 1:1 to 3:1
v) normally no provision for overflow
vi) plant growth is even

External catchment systems

The following are the main characteristics:

i) overland flow or rill flow harvested
ii) runoff stored in soil profile
iii) catchment usually 30 - 200 metres in length
iv) ratio catchment: cultivated area usually 2:1 to 10:1
v) provision for overflow of excess water
vi) uneven plant growth unless land levelled

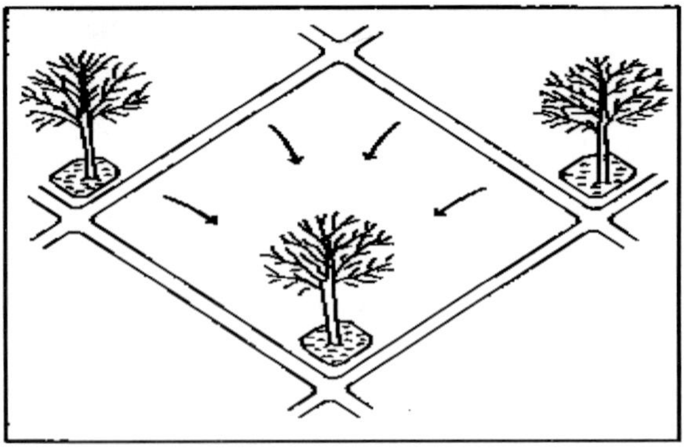

Figure 2. Microcatchment system: Negarim microcatchment for trees

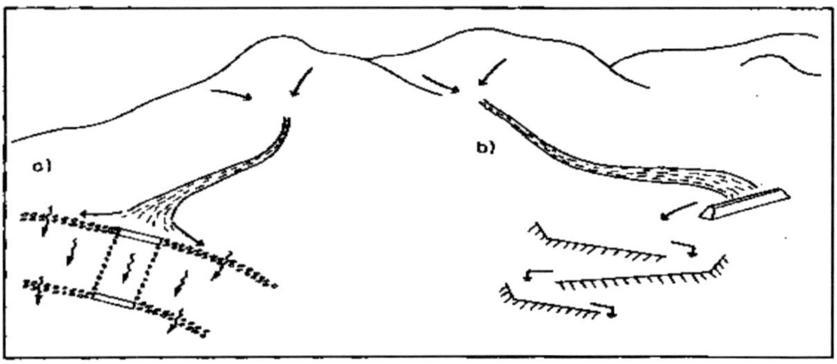

Figure 3. Floodwater farming systems: (a) spreading within channel bed; (b) diversion system

Floodwater farming

It often referred to as "Water Spreading" and sometimes "Spate Irrigation". The following are the main characteristics:

i) turbulent channel flow harvested either (a) by diversion or (b) by spreading within channel bed/valley floor
ii) runoff stored in soil profile
iii) catchment long (may be several kilometres)
iv) ratio catchment: cultivated area above 10:1
v) provision for overflow of excess water

Water Harvesting: Past and Future

Water harvesting is applied in arid and semi-arid regions where rainfall is either not sufficient to sustain a good crop and pasture growth or where, due to the erratic nature of precipitation, the risk of crop failure is very high. Water harvesting can significantly increase plant production in drought prone areas by concentrating the rainfall/runoff in parts of the total area. The intermittent character of rainfall and runoff and the ephemerality of floodwater flow requires some kind of storage. There might be some kind of interim storage in tanks, cisterns or reservoirs or soil itself serves as a reservoir for a certain period of time.

Water harvesting is based on the utilisation of surface runoff; therefore it requires runoff producing and runoff receiving areas. In most cases, with the exception of floodwater harvesting from far away catchments, water harvesting utilises the rainfall from the same location or region. Water harvesting projects are generally local and small scale projects.

The central elements of all water harvesting techniques are: a runoff area (catchment) with a sufficiently high run-off coefficient (impermeability would be optimal), and a "run-on" area, where the accumulated water is stored and/ or utilised. In most cases the runoff is used for agricultural crops, the water then being stored in the soil profile. A high storage capacity of the soil (i.e. medium textured soils) and a sufficient soil depth (> 1 m) are prerequisites here. The water retention capacity has to be high enough to supply the crops with water until the next rainfall event.

The most important parameters to be taken into consideration in practising water harvesting are therefore: rainfall distribution, rainfall intensity, runoff characteristics of the catchment, water storage capacity of soils, cisterns or reservoirs, the agricultural crops, available technologies and socio-economic conditions.

Water harvesting played a more important role in the past for the well-being of people in dry areas than it currently does. The reasons are manyfold:

i) no pumping from groundwater or other deep water sources

ii) very few large dams

iii) no long distance conveying of water through lined canals, pipes etc.

The building of structures for water harvesting, the cleaning and smoothing of runoff surfaces, the maintenance of canals and reservoirs etc. are labour demanding: Labour was a cheap resource, or even unpaid as in the case of slaves.

Agriculture was the backbone of the society and very few other choices to generate income were given. Therefore, relatively more input was invested in agriculture including runoff agriculture. Various examples shall be given to illustrate the past role of water harvesting worldwide.

In Jordan, there is indication of early water harvesting structures believed to have been constructed over 9,000 years ago. Evidence exists that simple water harvesting structures were used in Southern Mesopotamia as early as in 4,500 BC. Internationally, the most widely known runoff-irrigation systems have been found in the semi-arid to arid Negev desert region of Israel. Runoff agriculture in this region can be traced back as far as the 10th century BC when it was introduced by the Israelites of that period.

The Negev's most productive period in history however, began with the arrival of the Nabateans late in the 3rd century B.C. Runoff farming continued throughout Roman rule and reached its peak during the Byzantine era.

In North Yemen, a system dating back to at least 1,000 B.C. diverted enough floodwater to irrigate 20,000 hectares (50,000 acres) producing agricultural products that may have fed as many as 300,000 people. Farmers in this same area are still irrigating with floodwater, making the region perhaps one of the few places on earth where runoff agriculture has been continuously used since the earliest settlement.

In the South Tihama of Saudi Arabia, flood irrigation is traditionally used for sorghum production. Today, approximately 35,000 ha land, supporting 8,500 to 10,000 farm holdings, are still being flood irrigated.

In Baluchistan two water harvesting techniques were already applied in ancient times: the "Khuskaba" system and the "Sailaba" system. The first one employs bunds being built across the slope of the land to increase infiltration. The latter one utilises floods in natural water courses which are captured by earthen bunds.

In India, the "tank" system is traditionally the backbone of agricultural production in arid and semi-arid areas. The tanks collect rainwater and are constructed either by bunding or by excavating the ground. It is estimated that 4 to 10 hectares of catchment are required to fill one hectare of tank bed. In West Rajastan, with desert-like conditions having only 167 mm annual precipitation, large bunds were constructed as early as the 15th century to accumulate runoff. These "Khadin" create a reservoir which can be emptied at the end of the monsoon season to cultivate wheat and chickpeas with the remaining moisture.

A similar system called "Ahar" developed in the state of Bihar. Ahars are often built in series. It was observed that brackish groundwater in the neighbourhood of Ahars became potable after the Ahar was built, due to increased supply of rain water.

A very old flood diversion technique called "warping" is found in China's loess areas which harvests water as well as sediment.

Since at least Roman times water harvesting techniques were applied extensively in North Africa. Archeological research by the UNESCO Libyan Valleys team revealed that the wealth of the "granary of the Roman empire" was largely based on runoff irrigation. The team excavated structures in an area several hundred kilometres from the coast in the Libyan pre-desert, where the mean annual precipitation is well below fifty millimetres. The farming system here lasted well over 400 years and it sustained a large stationary population, often wealthy, which created enough crops to generate even a surplus. It produced barley, wheat, olive oil, grapes, figs, dates, sheep, cattle and pigs.

The precipitation is variable, falling in just one or two rain storms, often separated by droughts several years long. In Algeria, the "lacs collinaires", the rainwater storage ponds are traditional means of water harvesting for agriculture. The open ponds are mainly used for watering animals. In Tunesia, the "Meskat" and the "Jessour" systems have a long tradition, but are also still practised. The "Meskat" microcatchment system consists of an impluvium called "meskat", of about 500 m^2 in size, and a "manka" or cropping area of about 250 m^2. Thus, the CCR is 2:1. Both are surrounded by a 20 cm high bund, equipped with spillways to let runoff flow into the "manka" plots.

Whereas the "Meskats" are mainly found in the Sousse region, the "Jessour" are widespread in the South. The "Jessour" system is a terraced wadi system with earth dikes ("tabia") which are often reinforced by dry stone walls ("sirra"). The sediments accumulating behind the dikes are used for cropping.

In Lybia, archeological and historical studies have revealed the development and expansion of a highly successful dry (runoff based) farming agriculture during Roman times. On the slopes of the western and eastern mountain ranges some of these techniques continue to be practised.

In Egypt, the North-West coast and the Northern Sinai areas have a long tradition in water harvesting. Remnants from Roman times are frequently found. Some wadi terracing structures have been in use for over centuries.

Traditional techniques of water harvesting have been reported from many regions of Sub Saharan Africa. The central rangelands of Somalia are home to two small scale water harvesting systems which have been important local components of the production system for generations: The Caag system is a technique used to impound runoff from small water courses, gullies or even roadside drains. Sometimes ditches are dug to direct water into the fields. Runoff is impounded by the use of earth bunds. The entire plot may be a hectare or more in size. The alignment of the bunds is achieved by eye and by experience. In this system, runoff is impounded to a maximum depth of 30 cm. If water stands for more than five days or so, the bund may be deliberately breached to prevent waterlogging

The Mossi in Burkina Faso also constructed rock bunds and stone terraces in the past. The "Zay" system in Burkina Faso is a form of pitting which consists of digging holes that have a depth of 5 - 15 cm and a diameter of 10 - 30 cm. The usual spacing is between 50 - 100 cm. This results in a CCR of about 1 - 3:1. Manure and grasses are mixed with some of the soil and put into the zay. The rest of the soil is used to form a small dike downslope of the pit. Zay are applied in combination with bunds to conserve runoff, which is slowed down by the bunds. Many other traditional water harvesting systems existed or still exist, but the basic problem is that knowledge and information in this zone is extremely limited and fragmentary.

During this century, only very few water harvesting activities in research or implementation were undertaken before 1950. Australian farmers had already started to harvest water for domestic and animal use after World War I. During World War II, some water harvesting activities were carried out on islands with high rainfall.

In Jordan, earth dams have been constructed since 1964 in order to force runoff to infiltrate for pasture improvement. At the final stage the total area flooded shall be about 2,500 ha. In 1972, a project known as "Jordan Highland Development Project" was initiated.

Rock dams, contour stone bunds, trapezoidal bunds and earth contour bunds are used to increase soil moisture around the trees planted on steep lands. The total area utilised since its inception is estimated to be 6,000 hectares. Between 1985 and 1988, Jordan's Ministry of Agriculture, in collaboration with ACSAD, used contour terraces and ridges for pasture and range improvement in the Balama district. Better growth of olive, almond and pistachio was recorded on the experimental site.

In the Dei-Atiye community of Syria, rainwater harvesting was established in 1987 on an area of 130 ha. The project site was subdivided into four parts for tree crops, range plants, cereals and runoff research. The International Centre for Agricultural Research in the Dry Areas (ICARDA) in Syria, is currently working on the improvement of various WH techniques and on the identification of waterharvesting areas suitable for various West Asian and North African (WANA) environments.

In the North-West Arabia, a local system known as "Mahafurs" is still in use. This system is simply a shallow excavation of 20 - 100 m in diameter surrounded on three sides by earthen bunds 1 - 4 m high. The open side is pointed in the direction of water flow inside the wadi bed and used to collect water for animal consumption and moisture for plant production. In Afghanistan, composite microcatchments have been in use for a long time. In a survey conducted in the early 1970s, over 70,000 ha of Meskat-type systems used for growing fruit trees were reported.

In tropical Asia, especially southern India and Sri Lanka, earth bunds and excavated hollows have been used for runoff retention during the rainy seasons for millenia. Tank storage permits farmers to grow a second dry season irrigated crop in addition to rainfed agriculture. Tanks are sited randomly so that any overland flow from one is caught by others downslope. In case of siltation, labourers are hired to remove the silts and spread them on the cropping land during the dry season.

In different parts of Libya, experimental sites of contour-ridge terracing covering more than 53,000 ha have recently been established. In 1990, the government of Tunisia started the implementation of the National Strategy of Surface Runoff Mobilisation which aims, among other things, at building 21 dams, 203 small earth dams, 1,000 ponds, 2,000 works to recharge water tables and 2,000 works for irrigation through water spreading by the year 2,000.

In Wadi El-Arish region of Egypt, stone dykes were used to direct the runoff water flow for irrigation purposes. Also cisterns, which store water meant for animal and human consumption as well as for supplemental irrigation, are common in Egypt. The number of cisterns has increased from less than 3,000 in 1960 to about 15,000 in 1993 with a capacity of about 4 million m^3. In the North-West region of Egypt a GTZ/FAO sponsored project on land use planning including water harvesting activities was carried out (El-Shafey).

In Yemen, small dams storing runoff for later use in irrigation or rural supply have been constructed since the beginning of the eighties; the total storage capacity is between 50,000 to 90,000 m^3. "Matfia" is an old technique of storing water for human and animal consumption in Morocco, which still continues. Expensive modern technology, including the use of reinforced concrete has now been introduced in constructing the cisterns, although the local people are less interested in these large and expensive systems.

Since 1984, Morocco has started constructing dams to harvest flood water. The upstream catchment area under these dams ranges from 500 to 10,000 hectares. As of 1988, thirty five of these dams had been constructed. They provide irrigation water for about 160,000 animals and 3,000 ha of cultivated plots.

In Ethiopia, the Sudan and Botswana, small check-dams made of earth are used to catch moderate overland flow passing down slight slopes. They are called "haffirs" and support crops planted upslope.

In 1988, special ploughs developed by Italian scientists were used in Niger for the implementation of microcatchments on a large scale. Results from these plots showed excellent rates of tree establishment. A set of test plots on improved trapezoidal bunds in Baringo, Kenya has been constructed. This improved version consists of earth bunds which surround the plot and diverge as collection arms upslope to increase the catchment area. In 1984, a self-help project sponsored by Oxfam and known as Turkana Water Harvesting Project, was started in the Turkana district of Kenya. It was aimed at developing systems of water harvesting for crop production, while also introducing animal ploughing.

In Western Australia, topography modification in the form of catchment treatment has been practised for a long time. These are known as "roaded" catchments. They consist of parallel ridges ("roads") of steep, bare and compacted earth, surveyed at a gradient that allows runoff to occur without causing erosion of the intervening channels. In 1980, it was estimated that there were more than 3,500 roaded catchment systems in Western Australia, and many of them have a top dressing or a layer of compacted clay to increase the runoff efficiency.

Numerous water harvesting projects have failed because the technology used proved to be unsuitable for the specific conditions of the site. Each of the water harvesting methods has its advantages and limitations which can be summarised as follows: (1) Water harvesting for animal consumption In developing countries, the building or reactivation of cisterns and other rainwater tanks for animal consumption can save water which othewise has to be lifted or pumped from groundwater or carried over long distances.

During recent years some technological developments took place in regard to water harvesting which might have some impact on the future role of WH in general:

i) *Supplemental water system*: Runoff water is collected and stored offside for later application to the cropped area using some irrigation method. The water stored allows a prolongation of the cropping season or a second crop.

ii) *Dual purpose systems*: In a dual purpose system the runoff water flows first through the crop area, then the excess water is stored in some facility for later irrigation use. In Arizona, USA, runoff irrigation was combined e. g. with trickle irrigation, using sealed soil surfaces to increase runoff rates.

iii) *Combined systems*: If the irrigation water from aquifers or from rivers/reservoirs is not sufficient for year-round irrigation, a combination with runoff-irrigation (during the rainy season) is feasible. The combination of runoff-and furrow irrigation is reported from North Central Mexico.

iv) *Modelling*: If more information on hydrological, soil and crop parameters is available, models can be developed and applied to water harvesting for certain environments.

References

Abdi, A. M., "Water harvesting systems in the Northwestern region of Somalia", *Paper presented to the World Bank Workshop on Water Harvesting in Sub-Saharan Africa*, Baringo, Kenya, 13-17. Oct 1986.

Achouri, M., "Small Scale Water Harvesting in Tunisia", *In: FAO, Water Harvesting For Improved Agricultural Production*, Expert Consultation, Cairo, Egypt 21-25 Nov. 1993, p. 87-97, FAO, Rome, 1994.

Al-Labadi, M., "Water Harvesting in Jordan- Existing and Potential Systems", *In: FAO, Water Harvesting For Improved Agricultural Production*, Expert Consultation, Cairo, Egypt 21-25 Nov. 1993.

Barrow, C., *Water Resources and Agricultural Development in the Tropics*, Longman, UK, 1987.

Bruins, H. J., Evenari, M. and Nessler, U., *Rainwater harvesting agriculture for food production in arid zones: The challenge of the African famine*, Appl. Geogr. 6:13-33, 1986.

Huibers, F. P., "Rainfed Agriculture in a Semi-Arid Tropical Climate: Aspects of Land and Water Management for Red Soils in India", *Doctoral Thesis*, Agric. Univ. Wageningen, NL, 1985.

Oweis, T. and Taimeh, A., *Overall Evaluation of On-Farm Water the Arid Regions*, In: Lacirignola, C. and A. Hamdy (eds) Proceedings, CIHEAM Conference "Land and Water Resources Management in the Mediterranean Region" 04. - 08, September, 1994

Papadopoulos, I., "Water Harvesting from Greenhouses", *In: FAO, Water Harvesting For Improved Agricultural Production*, Expert Consultation, Cairo, Egypt 21-25 Nov. 1993, p. 97-101, FAO, Rome.

Prinz, D., Tauer, W., Voegtle, "The Application of Geographic Information Systems to Identify Areas Suitable for Water Harvesting/Runoff Irrigation" *Proceedings, XII CIGR World Congress*, Milano 29.08 - 01.09.1994.

........., "Water Harvesting: Past and Future", 1996.

Shatta, A. and F. Attia, "Environmental Aspects of Water Harvesting", *In: FAO, Water Harvesting For Improved Agricultural Production. Expert Consultation*, Cairo, Egypt 21-25 Nov. 1993.

Siegert, K., "Introduction to Water Harvesting. Some Basic Principles for Planning", *Design and Monitoring. In: FAO, Water Harvesting For Improved Agricultural Production*, Expert Consultation, Cairo, Egypt 21-25 Nov. 1993.

CHAPTER 2

Water and Soil Requirements

For the design of water harvesting systems, it is necessary to assess the water requirement of the crop intended to be grown. There have been various methods developed to determine the water requirement for specific plants. In the absence of any measured climatic data, it is often adequate to use estimates of water requirements for common crops (Table 1).

Table 1. Approximate values of seasonal crop water needs

Crop	Crop water need (mm/total growing period)
Beans	300 - 500
Citrus	900 - 1200
Cotton	700 - 1300
Groundnut	500 - 700
Maize	500 - 800
Sorghum/millet	450 - 650
Soybean	450 - 700
Sunflower	600 - 1000

Influencing Factors in Crop Water Requirements

Climate

A certain crop grown in a sunny and hot climate needs more water per day than the same crop grown in a cloudy and cooler climate. There are, however, apart from sunshine and temperature, other climatic factors which influence the crop water need. These factors are humidity and wind speed. When it is dry, the crop water needs are higher than when it is humid. In windy climates, the crops will use more water than in calm climates. The highest crop water needs are thus found in areas which are hot, dry, windy and sunny. The lowest values are found when it is cool, humid and cloudy with little or no wind.

From the above, it is clear that the crop grown in different climatic zones will have different water needs. For example, a certain maize variety grown in a cool climate will need less water per day than the same maize variety grown in a hotter climate.

It is therefore useful to take a certain standard crop or reference crop and determine how much water this crop needs per day in the various climatic regions. As a standard crop (or reference crop) grass has been chosen. Table 3 indicates the average daily water needs of this reference grass crop. The daily water needs of the grass depend on the climatic zone (rainfall regime) and daily temperatures.

Table 2. Effect of major climatic factors on crop water needs

Climatic factor	Crop water need	
	High	Low
Sunshine	sunny (no clouds)	cloudy (no sun)
Temperature	hot	cool
Humidity	low (dry)	high (humid)
Wind speed	windy	little wind

Table 3. Average daily water need of standard grass during irrigation season (mm)

Climatic zone	Mean daily temperature		
	low (< 15°C)	medium (15-25°C)	high (> 25°C)
Desert/arid	4-6	7-8	9-10
Semi-arid	4-5	6-7	8-9

For the various field crops it is possible to determine how much water they need compared to the standard grass. A number of crops need less water than grass, a number of crops need more water than grass and other crops need more or less the same amount of water as grass. Understanding of this relationship is extremely important for the selection of crops to be grown in a water harvesting scheme.

Table 4. Crop water needs in peak period of various crops compared to the standard grass crop

-30%	-10%	same as standard grass	+10%	+20%
Citrus	Squash	Crucifers	Barley	
Wheat				
Wheat		Nuts& Olives	Beans	fruit
trees				
Groundnuts	Maize			cover
		Melons Onions Peppers Grass Clean cultivated nuts & fruit trees	Cotton Lentils Millet	
			Sorghum Soybeans Sunflower	

Crop type on Crop water needs

The influence of the crop type on the crop water need is important in two ways.

a) The crop type has an influence on the daily water needs of a fully grown crop; i.e. the peak daily water needs of a fully developed maize crop will need more water per day than a fully developed crop of onions.
b) The crop type has an influence on the duration of the total growing season of the crop. There are short duration crops, e.g. peas, with a duration of the total growing season of 90-100 days and longer duration crops, e.g. melons, with a duration of the total growing season of 120-160 days. There are, of course, also perennial crops that are in the field for many years, such as fruit trees.

While, for example, the daily water need of melons may be less than the daily water need of beans, the seasonal water need of melons will be higher than that of beans because the duration of the total growing season of melons is much longer. Data on the duration of the total growing season of the various crops grown in an area can best be obtained locally. These data may be obtained from, for example, the seed supplier, the Extension Service, the Irrigation Department or Ministry of Agriculture.

Table 5 gives some indicative values or approximate values for the duration of the total growing season for the various field crops. It should, however, be noted that the values are only rough approximations and it is much better to obtain the values locally.

Table 5. Indicative Values of total growing period

Crop	Total growing period (days)	Crop	Total growing period (days)
Alfalfa	100-365	Melon	120-160
Barley/Oats/ Wheat	120-150	Millet	105-140
Bean, green	75-90	Onion, green	70-95
dry	95-110	dry	150-210
Citrus	240-365	Pepper	120-210
Cotton	180-195	Rice	90-150
Grain/small	150-165	Sorghum	120-130
Lentil	150-170	Soybean	135-150
Maize, sweet	80-110	Squash	95-120
grain	125-180	Sunflower	125-130

From Table 5, it is obvious that there is a large variation of values not only between crops, but also within one crop type. In general, it can be assumed that the growing period for a certain crop is longer when the climate is cool and shorter when the climate is warm. Crops differ in their response to moisture deficit. This characteristic is commonly termed "drought resistance". When crop water requirements are not met, crops with a high drought sensitivity suffer greater reductions in yields than crops with a low sensitivity.

Table 6. General Sensitivity to drought

Group One:	(low sensitivity)	Groundnuts
	–	Safflower
Group Two:	–	Sorghum
	–	Cotton
	–	Sunflower
Group Three:	–	Beans
Group Four:	(high sensitivity)	Maize

Methods of calculation of crop water requirements

The calculation of crop water requirements by means of the two methods. The basic formula for the calculation reads as follows:

$$ETcrop = kc \times Eto$$

where:

ETcrop = the water requirement of a given crop in mm per unit of time e.g. mm/day, mm/month or mm/season.

kc = the "crop factor"

ETo = the "reference crop evapotranspiration" in mm per unit of time e.g. mm/day, mm/month or mm/season.

ETo – Reference Crop Evapotranspiration

The reference crop evapotranspiration ETo (sometimes called potential evapotranspiration, PET) is defined as the rate of evapotranspiration from a large area covered by green grass which grows actively, completely shades the ground and which is not short of water. The rate of water which evapotranspirates depends on the climate. The highest value of ETo is found in areas which are hot, dry, windy and sunny whereas the lowest values are observed in areas where it is cool, humid and cloudy with little or no wind.

In many cases it will be possible to obtain estimates of ETo for the area of concern (or an area nearby with similar climatic conditions) from the Meteorological Service. However, where this is not possible, the values for ETo have to be calculated. Two easy methods will be explained below:

Pan evaporation method

With this method, ETo can be obtained by using evaporation rates which are directly measured with an evaporation pan. This is a shallow pan, containing water which is exposed to the evaporative influence of the climate. The standard pan is shown in Figure 1. It has a diameter of 1.21 m, a depth of 25 cm and is placed 15 cm above the ground.

Figure 1. Class A evaporation pan

An evaporation pan is easy to construct and in most situations the material can be found locally. The principle of obtaining evaporation rates from the pan is as follows:

i) the pan is installed in the field 15 cm above the ground;

ii) the pan is filled with water 5 cm below the rim;

iii) the water is allowed to evaporate during a certain period of time (usually 24 hours). For example, each morning at 7.00 hours a measurement is taken. Rainfall, if any, is measured simultaneously;

iv) after 24 hours, the water depth is measured again;

v) the amount of water which has evaporated in a given time unit is equal to the difference between the two measured water depths. This is the pan evaporation rate: Epan (mm/24 hours).

The readings taken from the pan (Epan) however do not give ETo directly, but have to be multiplied by a "Pan Coefficient" (Kpan). Thus:

$$ETo = Epan ´ Kpan$$

For the Class A evaporation pan, Kpan varies between 0.35 and 0.85, with an average of 0.70. If the precise pan factor is not known, the average value (0.70) can be used as an approximation.

The Blaney-Criddle Method

If no measured data on pan evaporation are available, the Blaney-Criddle method can be used to calculate ETo. This method is straightforward and requires only data on mean daily temperatures. However, with this method, only approximations of ETo are obtained which can be inaccurate in extreme conditions.

The Blaney-Criddle formula is: ETo = p(0.46Tmean + 8) where:

ETo = reference crop evapotranspiration (mm/day)

Tmean = mean daily temperature (°C)

p = mean daily percentage of annual daytime hours.

The Blaney-Criddle Method always refers to mean monthly values, both for the temperature and the ETo. If in a local meteorological station the daily minimum and maximum temperatures are measured, the mean daily temperature is calculated as follows:

$$Tmin = \frac{\text{sum of all T min values during the month}}{\text{number of days of the month}}$$

To determine the value of p. Table 8 is used. To be able to obtain the p value, it is essential to know the approximate latitude of the area: the number of degrees north or south of the Equator.

Table 7. Mean daily percentage (p) of annual daytime hours for different latitudes

Latitude:												
N.	Jan	Feb	Mar	Apr	May	Jun	July	Aug	Sept	Oct	Nov	Dec
S.	July	Aug	Sept	Oct	Nov	Dec	Jan	Feb	Mar	Ap	May	June
60°	.15	.20	.26	.32	.38	.41	.40	.34	.28	.22	.17	.13
55	.17	.21	.26	.32	.36	.39	.38	.33	.28	.23	.18	.16
50	.19	.23	.27	.31	.34	.36	.35	.32	.28	.24	.20	.18
45	.20	.23	.27	.30	.34	.35	.34	.32	.28	.24	.21	.20
40	.22	.24	.27	.30	.32	.34	.33	.31	.28	.25	.22	.21
35	.23	.25	.27	.29	.31	.32	.32	.30	.28	.25	.23	.22
30	.24	.25	.27	.29	.31	.32	.31	.30	.28	.26	.24	.23
25	.24	.26	.27	.29	.30	.31	.31	.29	.28	.26	.25	.24
20	.25	.26	.27	.28	.29	.30	.30	.29	.28	.26	.25	.25
15	.26	.26	.27	.28	.29	.29	.29	.28	.28	.27	.26	.25
10	.26	.27	.27	.28	.28	.29	.29	.28	.28	.27	.26	.26
5	.27	.27	.27	.28	.28	.28	.28	.28	.28	.27	.27	.27
0	.27	.27	.27	.27	.27	.27	.27	.27	.27	.27	.27	.27

For example, when p = 0.29 and T mean = 21.5°C, the ETo is calculated as follows: ETo = 0.29 (0.46 ´ 21.5 + 8) = 0.29 (9.89 + 8) = 0.29 ´ 17.89 = 5.2 mm/day.

Table 8. Indicative values of Eto (mm/day)

Climatic zone	Mean daily temperature		
	15°	15-25°C	25°
Desert/arid	4-6	7-8	9-10
Semi-arid	4-5	6-7	8-9
Sub-humid	3-4	5-6	7-8
Humid	1-2	3-4	5-6

Indicative values of ETo

Table 9 contains approximate values fur ETo which may be used in the absence of measured or calculated figures.

Crop Factor - Kc

In order to obtain the crop water requirement ETcrop the reference crop evapotranspiration, ETo, must be multiplied by the crop factor, Kc. The crop factor (or "crop coefficient") varies according to the growth stage of the crop. There are four growth stages to distinguish:

i) the initial stage: when the crop uses little water;
ii) the crop development stage, when the water consumption increases;
iii) the mid-season stage, when water consumption reaches a peak;
iv) the late-season stage, when the maturing crop once again requires less water.
v) Calculation of ETcrop

While conventional irrigation strives to maximise the crop yields by applying the optimal amount of water required by the crops at well determined intervals, this is not possible with water harvesting techniques.

Water Requirements of Multipurpose Trees

There is little information available about the water requirements of multipurpose trees planted under rainwater harvesting systems in semi-arid areas. In general, the water requirements for trees are more difficult to determine than for crops. Trees are relatively sensitive to moisture stress during the establishment stage compared with their ability to withstand drought once their root systems are fully developed. There is no accurate information available on the response of these species, in terms of yields, to different irrigation/water regimes.

Table 9 gives some basic data of multipurpose trees often planted in semi-arid areas. The critical stage for most trees is in the first two years of seedling/sapling establishment.

Table 9. Naturally referred climatic zones of multipurpose trees

	Semi-arid/marginal 500-900 mm rain	Arid/semi-arid 150-500 mm	Tolerance to rain temporary waterlogging
Acacia albida	yes	yes	yes
A. nilotica	yes	yes	yes
A. saligna	no	yes	yes
A. senegal	yes	yes	no
A. seyal	yes	yes	yes
A. tortilis	yes	yes	no
Albizia lebbeck	yes	no	no
Azadirachta indica	yes	no	some
Balanites aegyptiaca	yes	yes	yes
Cassia siamea	yes	no	no
Casuarina equisetifolia	yes	no	some
Colophospermum mopane	yes	yes	yes
Cordeauxia edulis	no	yes	?
Cordia sinensis	no	yes	?
Delonix elata	yes	no	?
Eucalyptus camaldulensis	yes	yes	yes
Prosopis chilensis	yes	yes	some
Prosopis cineraria	yes	yes	yes
Prosopis juliflora	yes	yes	yes
Ziziphus mauritiana	yes	yes	yes

Water Requirements for Fruit trees

There are some known values of water requirements for fruit trees under water harvesting systems. Table 10 contains the water requirements for some fruit trees.

Table 10. Fruit tree water requirements

Species	Seasonal water requirement	Place	
Apricots	550 mm*	Israel	
Peaches	700 mm*	Israel	
	Pomegranate	265 mm	Israel
Jujube (Zixiphus mauritiana)	550-750 mm	India	

** This figure is the full irrigation rate. Where there was no irrigation but only rainwater harvesting the equivalent of 270 mm depth was adequate to support the trees.*

Water Requirements of Rangeland and Fodder

Water requirements for rangeland and fodder species grown in semi-arid/arid areas under WH systems are usually not calculated. The objective is to improve performance, within economic constraints, and to ensure the survival of the plants from season to season, rather than fully satisfying water requirements.

Soil Requirements for Water Harvesting

Generally the soil characteristics for water harvesting should be the same as those for irrigation. Ideally the soil in the catchment area should have a high runoff coefficient while the soil in the cultivated area should be a deep, fertile loam. Where the conditions for the cultivated and catchment areas conflict, the requirements of the cultivated area should always take precedence. The following are important aspects of soils which affect plant performance under WH systems.

Texture: The texture of a soil has an influence on several important soil characteristics including infiltration rate and available water capacity. Soil texture refers to its composition in terms of mineral particles. A broad classification is:

a. Coarse textured soils - sand predominant - "sandy soils"

b. Medium textured soils - silt predominant - "loamy soils"

c. Fine textured soils - clay predominant - "clayey soils"

Generally speaking it is the medium textured soils, the loams, which are best suited to WH system since these are ideally suited for plant growth in terms of nutrient supply, biological activity and nutrient and water holding capacities.

Structure: Soil structure refers to the grouping of soil particles into aggregates, and the arrangement of these aggregates. A good soil structure is usually associated with loamy soil and a relatively high content of organic matter. Inevitably, under hot climatic conditions, organic matter levels are often low, due to the rapid rates of decomposition. The application of organic materials such as crop residues and animal manure is helpful in improving the structure.

Depth: The depth of soil is particularly important where WH systems are proposed. Deep soils have the capacity to store the harvested runoff as well as providing a greater amount of total nutrients for plant growth. Soils of less than one metre deep are poorly suited to WH. Two metres depth or more is ideal, though rarely found in practice.

Fertility: In many of the areas where WH systems may be introduced, lack of moisture and low soil fertility are the major constraints to plant growth. Some areas in Sub-Saharan Africa, for example, may be limited by low soil fertility as much as by lack of moisture. Nitrogen and phosphorus are usually the elements most deficient in these soils. While it is often not possible to avoid poor soils in areas under WH system development, attention should be given to the maintenance of fertility levels.

Salinity/sodicity: Sodic soils, which have a high exchangeable sodium percentage, and saline soil which have excess soluble salts, should be avoided for WH systems. These soils can reduce moisture availability directly, or indirectly, as well as exerting direct harmful influence on plant growth.

Infiltration rate: The infiltration rate of a soil depends primarily on its texture. Typical comparative figures of infiltration are as follows:

	mm/hour
sandy soil	50
sandy loam	25
loam 12.5	
clay loam	7.5

A very low infiltration rate can be detrimental to WH systems because of the possibility of waterlogging in the cultivated area. On the other hand, a low infiltration rate leads to high runoff, which is desirable for the catchment area. The soils of the cropped area however should be sufficiently permeable to allow adequate moisture to the crop root zone without causing waterlogging problems. Therefore, the requirements of the cultivated area should always take precedence. Crust formation is a special problem of arid and semi-arid areas, leading to high runoff and low infiltration rates. Soil compaction as a result of heavy traffic either from machinery or grazing animals could also result in lower infiltration rates.

Available water capacity (AWC): The capacity of soils to hold, and to release adequate levels of moisture to plants is vital to WH. AWC is a measure of this parameter, and is expressed as the depth of water in mm readily available to plants after a soil has been thoroughly wetted to "field capacity". AWC values for loams vary from 100-200 mm/metre. Not only is the AWC important, but the depth of the soil is critical also. In WH systems which pond runoff, it is vital that this water can be held by the soil and made available to the plants.

The AWC has implications for technical design - for example simple calculation demonstrates that even in deep soils (2 metres) with high AWC values (200 mm/metre) there is no point ponding water to depths greater than 40 cm. This quantity when infiltrated is adequate to replenish the soil profile from permanent wilting point to field capacity and any surplus will be lost by deep drainage as well as being a potential waterlogging hazard.

References

Tayaa, M., "Present Situation and Prospects for Improvement of Water Harvesting Practices in Morocco", *In: FAO, Water Harvesting For Improved Agricultural Production*, Expert Consultation, Cairo, Egypt 21-25 Nov. 1993.

Tobbi, B., "Water Harvesting: Historic, Existing and Potentials in Tunisia", *In: FAO, Water Harvesting For Improved Agricultural Production, Expert Consultation*, Cairo, Egypt 21-25 Nov. 1993.

UNEP, *Rain and Stormwater Harvesting in Rural Areas*, Tycooly, Dublin, 1983.

Van Dijk, J. and C. Reij, "Indigenous Water Harvesting Techniques in Sub-Saharan Africa: Examples from Sudan and the West African Sahel", *In: FAO, Water Harvesting For Improved Agricultural Production*, Expert Consultation, Cairo, Egypt 21-25 Nov. 1993.

Vijayalakshmi, K., K. P. R. Vittal and L. R. P. Singh, "Water Harvesting and Re-use". *In: ICAR. A Decade of Dryland Agricultural Research in India 1971-1980*, p. 103-119. New Delhi, India, 1982.

Zohar, Y., J.A. Aronson and H.Loevenstein, "Fuelwood and Fodder Crops in Rainfall Harvesting Systems Under Arid Conditions" *A Research Progress Report*, ARO Volcani Center. Bet Dagan, Israel, 1987.

CHAPTER 3

Rainfall-runoff Analysis

Runoff is generated by rainstorms and its occurrence and quantity are dependent on the characteristics of the rainfall event, i.e. intensity, duration and distribution. There are, in addition, other important factors which influence the runoff generating process.

Characteristics of Rainfall

Precipitation in arid and semi-arid zones results largely from convective cloud mechanisms producing storms typically of short duration, relatively high intensity and limited areal extent. However, low intensity frontal-type rains are also experienced, usually in the winter season.

When most precipitation occurs during winter, as in Jordan and in the Negev, relatively low-intensity rainfall may represent the greater part of annual rainfall. Rainfall intensity is defined as the ratio of the total amount of rain (rainfall depth) falling during a given period to the duration of the period It is expressed in depth units per unit time, usually as mm per hour (mm/h).

The statistical characteristics of high-intensity, short-duration, convective rainfall are essentially independent of locations within a region and are similar in many parts of the world. Analysis of short-term rainfall data suggests that there is a reasonably stable relationship governing the intensity characteristics of this type of rainfall.

Annual Rainfall

Water harvesting planning and management in arid and semi-arid zones present difficulties which are due less to the limited amount of rainfall than to the inherent degree of variability associated with it. In temperate climates, the standard deviation of annual rainfall is about 10-20 percent and in 13 years out of 20, annual amounts are between 75 and 125 percent of the mean. In arid and semi-arid climates the ratio of maximum to minimum annual amounts is much greater and the annual rainfall distribution becomes increasingly skewed with increasing aridity. With mean annual rainfalls of 200-300 mm the rainfall in 19 years out of 20 typically ranges from 40 to 200 percent of the mean and for 100 mm/year, 30 to 350 percent of the mean. At more arid locations it is not uncommon to experience several consecutive years with no rainfall.

For a water harvesting planner, the most difficult task is therefore to select the appropriate "design" rainfall according to which the ratio of catchment to cultivated area will be determined. Design rainfall is defined as the total amount of rain during the cropping season at which or above which the catchment area will provide sufficient runoff to satisfy the crop water requirements. If the actual rainfall in the cropping season is below the design rainfall, there will be moisture stress in the plants; if the actual rainfall exceeds the design rainfall, there will be surplus runoff which may result in a damage to the structures.

Design Rainfall

The design rainfall is usually assigned to a certain probability of occurrence or exceedance. If, for example, the design rainfall with a 67 percent probability of exceedance is selected, this means that on average this value will be reached or exceeded in two years out of three and therefore the crop water requirements would also be met in two years out of three. The design rainfall is determined by means of a statistical probability analysis.

A rather simple, graphical method to determine the probability or frequency of occurrence of yearly or seasonal rainfall. The first step is to obtain annual rainfall totals for the cropping season from the area of concern. In locations where rainfall records do not exist, figures from stations nearby may be used with caution. It is important to obtain long-term records.

Rainfall-runoff Relationship

Surface Runoff Process

When rain falls, the first drops of water are intercepted by the leaves and stems of the vegetation. This is usually referred to as interception storage.

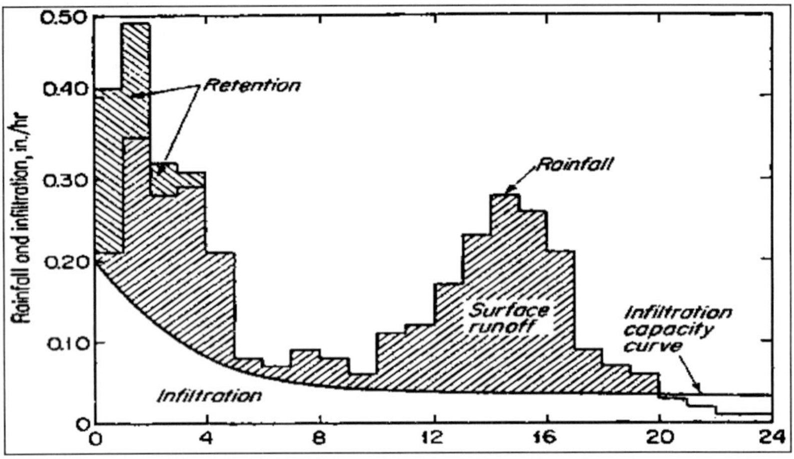

Figure 1. Schematic diagram illustrating relationship between rainfall, infiltration and runoff

As the rain continues, water reaching the ground surface infiltrates into the soil until it reaches a stage where the rate of rainfall (intensity) exceeds the infiltration capacity of the soil. Thereafter, surface puddles, ditches, and other depressions are filled (depression storage), after which runoff is generated. The infiltration capacity of the soil depends on its texture and structure, as well as on the antecedent soil moisture content (previous rainfall or dry season). The initial capacity (of a dry soil) is high but, as the storm continues, it decreases until it reaches a steady value termed as final infiltration rate.

The process of runoff generation continues as long as the rainfall intensity exceeds the actual infiltration capacity of the soil but it stops as soon as the rate of rainfall drops below the actual rate of infiltration. The rainfall runoff process is well described in the literature. Numerous papers on the subject have been published and many computer simulation models have been developed. For a better understanding of the difficulties of accurately predicting the amount of runoff resulting from a rainfall event, the major factors which influence the rainfall-runoff process are described below.

Apart from rainfall characteristics such as intensity, duration and distribution, there are a number of site (or catchment) specific factors which have a direct bearing on the occurrence and volume of runoff.

i. Soil type

The infiltration capacity is among others dependent on the porosity of a soil which determines the water storage capacity and affects the resistance of water to flow into deeper layers. Porosity differs from one soil type to the other. The highest infiltration capacities are observed in loose, sandy soils while heavy clay or loamy soils have considerable smaller infiltration capacities.

The infiltration capacity depends furthermore on the moisture content prevailing in a soil at the onset of a rainstorm. The initial high capacity decreases with time (provided the rain does not stop) until it reaches a constant value as the soil profile becomes saturated (Figures 1 and 2).

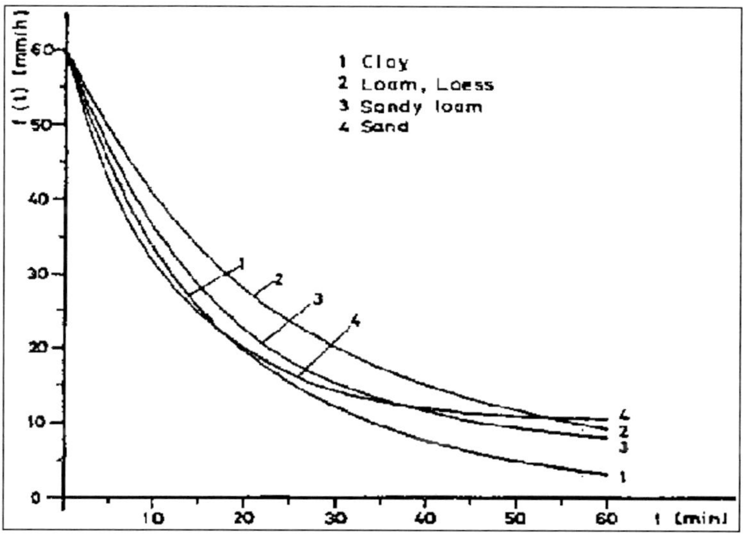

Figure 2. The difference in infiltration capacities measured in different soil types.

This however, is only valid when the soil surface remains undisturbed. It is well known that the average size of raindrops increases with the intensity of a rainstorm. In a high intensity storm the kinetic energy of raindrops is considerable when hitting the soil surface. This causes a breakdown of the soil aggregate as well as soil dispersion with the consequence of driving fine soil particles into the upper soil pores. This results in clogging of the pores, formation of a thin but dense and compacted layer at the surface which highly reduces the infiltration capacity.

This effect, often referred to as capping, crusting or sealing, explains why in arid and semi-arid areas where rainstorms with high intensities are frequent, considerable quantities of surface runoff are observed even when the rainfall duration is short and the rainfall depth is comparatively small. Soils with a high clay or loam content are the most sensitive for forming a cap with subsequently lower infiltration capacities. On coarse, sandy soils the capping effect is comparatively small.

ii. Vegetation

The amount of rain lost to interception storage on the foliage depends on the kind of vegetation and its growth stage. Values of interception are between 1 and 4 mm. A cereal crop, for example, has a smaller storage capacity than a dense grass cover. More significant is the effect the vegetation has on the infiltration capacity of the soil. A dense vegetation cover shields the soil from the raindrop impact and reduces the crusting effect as described earlier. In addition, the root system as well as organic matter in the soil increase the soil porosity thus allowing more water to infiltrate. Vegetation also retards the surface flow particularly on gentle slopes, giving the water more time to infiltrate and to evaporate.

iii.Slope and catchment size

Investigations on experimental runoff plots have shown that steep slope plots yield more runoff than those with gentle slopes. In addition, it was observed that the quantity of runoff decreased with increasing slope length. This is mainly due to lower flow velocities and subsequently a longer time of concentration (defined as the time needed for a drop of water to reach the outlet of a catchment from the most remote location in the catchment). This means that the water is exposed for a longer duration to infiltration and evaporation before it reaches the measuring point. The same applies when catchment areas of different sizes are compared. The runoff efficiency (volume of runoff per unit of area) increases with the decreasing size of the catchment i.e. the larger the size of the catchment the larger the time of concentration and the smaller the runoff efficiency. Figure 3 clearly illustrates this relationship.

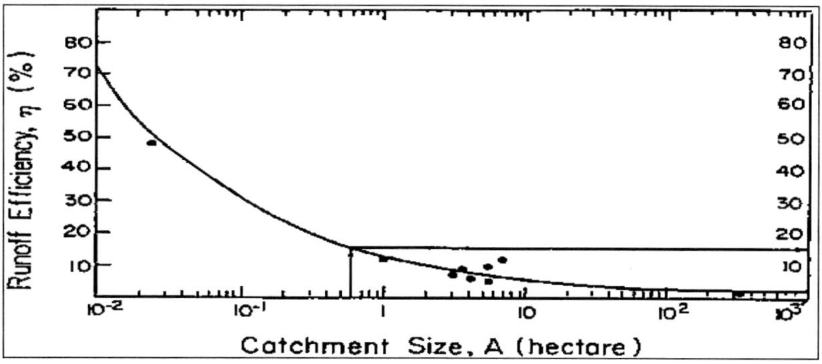

Figure 3. Runoff efficiency as a function of catchment size

The purpose of this diagram is to demonstrate the general trend between runoff and catchment size.

Runoff Coefficients

The site-specific factors which strongly influence the rainfall-runoff process, it should also be considered that the physical conditions of a catchment area are not homogenous. Even at the micro level there are a variety of different slopes, soil types, vegetation covers etc. Each catchment has therefore its own runoff response and will respond differently to different rainstorm events. The design of water harvesting schemes requires the knowledge of the quantity of runoff to be produced by rainstorms in a given catchment area. It is commonly assumed that the quantity of runoff is a proportion (percentage) of the rainfall depth.

Runoff [mm] = K ´ Rainfall depth [mm]

In rural catchments where no or only small parts of the area are impervious, the coefficient K, which describes the percentage of runoff resulting from a rainstorm, is however not a constant factor. Instead its value is highly variable and depends on the above described catchment-specific factors and on the rainstorm characteristics.

For example, in a specific catchment area with the same initial boundary condition (e.g. antecedent soil moisture), a rainstorm of 40 minutes duration with an average intensity of 30 mm/h would produce a smaller percentage of runoff than a rainstorm of only 20 minutes duration but with an average intensity of 60 mm/h although the total rainfall depth of both events were equal.

Determination of runoff coefficients

For reasons explained before, the use of runoff coefficients which have been derived for watersheds in other geographical locations should be avoided for the design of a water harvesting scheme. Also runoff coefficients for large watersheds should not be applied to small catchment areas. An analysis of the rainfall-runoff relationship and subsequently an assessment of relevant runoff coefficients should best be based on actual, simultaneous measurements of both rainfall and runoff in the project area. The runoff coefficient from an individual rainstorm is defined as runoff divided by the corresponding rainfall both expressed as depth over catchment area (mm):

$$K = \frac{\text{Runoff [mm]}}{\text{Rainfall [mm]}}$$

Studies shows that at least 2 years should be spent to measure rainfall and runoff data before any larger construction programme starts. Such a time span would in any case be justified bearing in mind the negative demonstration effect a water harvesting project would have if the structures were seriously damaged or destroyed already during the first rainstorm because the design was based on erroneous runoff coefficients. When plotting the runoff coefficients against the relevant rainfall depths a satisfactory correlation is usually observed.

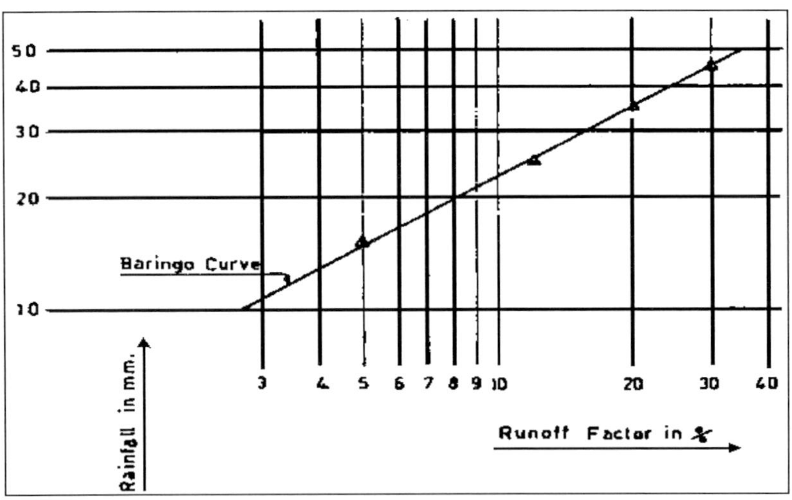

Figure 4. Rainfall-runoff relationships, Baringo, Kenya

A much better relationship would be obtained if in addition to rainfall depth the corresponding rainstorm intensity, the rainstorm duration and the antecedent soil moisture were also measured. This would allow rainstorm events to be grouped according to their average intensity and their antecedent soil moisture and to plot the runoff coefficients against the relevant rainfall durations separately for different intensities. Rainfall intensities can be accurately measured by means of a continuously recording autographic rain gauge. It is also possible to time the length of individual rainstorms and to calculate the average intensities by dividing the measured rainfall depths by the corresponding duration of the storms.

When analysing the measured data it will be noted that a certain amount of rainfall is always required before any runoff occurs. This amount, usually referred to as threshold rainfall, represents the initial losses due to interception and depression storage as well as to meet the initially high infiltration losses. The threshold rainfall depends on the physical characteristics of the area and varies from catchment to catchment.

In areas with only sparse vegetation and where the land is very regularly shaped, the threshold rainfall may be only in the range of 3 mm while in other catchments this value can easily exceed 12 mm, particularly where the prevailing soils have a high infiltration capacity. The fact that the threshold rainfall has first to be surpassed explains why not every rainstorm produces runoff. This is important to know when assessing the annual runoff-coefficient of a catchment area.

Assessment of Seasonal Runoff

The knowledge of runoff from individual storms as described before is essential to assess the runoff behaviour of a catchment area and to obtain an indication both of runoff-peaks which the structure of a water harvesting scheme must withstand and of the needed capacity for temporary surface storage of runoff, for example the size of an infiltration pit in a microcatchment system. The annual (seasonal) runoff coefficient differs from the runoff coefficients derived from individual storms as it takes into account also those rainfall events which did not produce any runoff. The annual (seasonal) runoff-coefficient is therefore always smaller than the arithmetic mean of runoff coefficients derived from individual runoff-producing storms.

Runoff plots

Runoff plots are used to measure surface runoff under controlled conditions. The plots should be established directly in the project area. Their physical characteristics, such as soil type, slope and vegetation must be representative of the sites where water harvesting schemes are planned. The size of a plot should ideally be as large as the estimated size of the catchment planned for the water harvesting project. This is not always possible mainly due to the problem of storing the accumulated runoff.

A minimum size of 3-4 m in width and 10-12 m in length is recommended. Smaller dimensions should be avoided, since the results obtained from very small plots are rather misleading. Care must be taken to avoid sites with special problems such as rills, cracks or gullies crossing the plot. These would drastically affect the results which would not be representative for the whole area. The gradient along the plot should be regular and free of local depressions.

During construction of the plot, care must be taken not to disturb or change the natural conditions of the plot such as destroying the vegetation or compacting the soil. It is advisable to construct several plots in series in the project area which would permit comparison of the measured runoff volumes and to judge on the representative character of the selected plot sites. Around the plots metal sheets or wooden planks must be driven into the soil with at least 15 cm of height above ground to stop water flowing from outside into the plot and vice versa.

A rain gauge must be installed near to the plot. At the lower end of the plot a gutter is required to collect the runoff. The gutter should have a gradient of 1% towards the collection tank. The soil around the gutter should be backfilled and compacted. The joint between the gutter and the lower side of the plot may be cemented to form an apron in order to allow a smooth flow of water from the plot into the gutter. The collection tank may be constructed from stone masonry, brick or concrete blocks, but a buried barrel will also meet the requirements. The tank should be covered and thus be protected against evaporation and rainfall.

The storage capacity of the tank depends on the size of the plot but should be large enough to collect water also from extreme rain storms. Following every storm (or every day at a specific time), the volume of water collected in the rain gauge and in the runoff tank must be measured. Thereafter the gauge and tank must be completely emptied. Any silt which may have deposited in the tank and in the gutter must be cleared.

References

Aminipouri, B & Ghoddousi, J., "Rainwater Catchment for Survival", *Proceedings of the 8th International Rainwater catchment Systems Conference*, Tehran: Ministry of Jihad-e-Sazandegi, 1997.

Cunliffe, D. "Guidance on the Use of Rainwater Tanks, National Environmental Health Forum

Monographs", *Water Series* 3, Public and Environmental Health Service, Dept. of Human Services, PO Box 6, Rundle Mall SA 5000, Australia, 1998.

Gleick, P.ed., *Water in Crisis: A guide to the World's Fresh Water Resources*, Oxford Univ.Press, 1993.

Gould, J. & Nissen-Petersen, E., *Rainwater Catchment Systems for Domestic Supply: Design, Construction and Implementation* , London: I.T. Publications, 1999.

König, K., *Rainwater in Cities, Ecological Concepts,* fbr, Kassler Str. 1a, D-60486, Frankfurt am Main, Germany, 1998.

Mou, H. et. al. Edit., "Rainwater Utilization for the World's People, Proceedings of the 7th International Rainwater Catchment Systems Conference", *Chinese Academy of Sciences*, Beijing, China, Volumes. 1 and 2., 1995.

Pacey, A. & Cullis, A., *Rainwater Harvesting: The Collection of Rainfall and Runoff in Rural Areas*, London: I.T. Publications, 1986.

Pearce, F., *The Dammed: Rivers, Dams and the coming world water crisis*, London: The Bodley Head, 1992.

Thomas, T., *Domestic Water Supply Using Rainwater Harvesting, Building Research and Information* Vol 2, No 2: 94-101, 1998.

CHAPTER 4

Water Harvesting Techniques

A water harvesting scheme will only be sustainable if it fits into the socio-economic context of the area and also fulfills a number of basic technical criteria.

Slopes: The ground slope is a key limiting factor to water harvesting. Water harvesting is not recommended for areas where slopes are greater than 5% due to uneven distribution of run-off and large quantities of earthwork required which is not economical.

Soils: Should have the main attributes of soils which are suitable for irrigation: they should be deep, not be saline or sodic and ideally possess inherent fertility. A serious limitation for the application of water harvesting are soils with a sandy texture. If the infiltration rate is higher than the rainfall intensity, no runoff will occur.

Costs: The quantities of earth/stonework involved in construction directly affects the cost of a scheme or, if it is implemented on a self help basis, indicates how labour intensive its construction will be.

Negarim Microcatchments

Negarim microcatchments are diamond-shaped basins surrounded by small earth bunds with an infiltration pit in the lowest corner of each. Runoff is collected from within the basin and stored in the infiltration pit. Microcatchments are mainly used for growing trees or bushes. This technique is appropriate for small-scale tree planting in any area which has a moisture deficit. Besides harvesting water for the trees, it simultaneously conserves soil. Negarim microcatchments are neat and precise, and relatively easy to construct.

Israel has the most widespread and best developed Negarim microcatchments, mostly located on research farms in the Negev Desert, where rainfall is as low as 100-150 mm per annum. However the technique, and variations of it, is widely used in other semi-arid and arid areas, especially in North and Sub-Saharan Africa Because it is a well-proven technique, it is often one of the first to be tested by new projects.

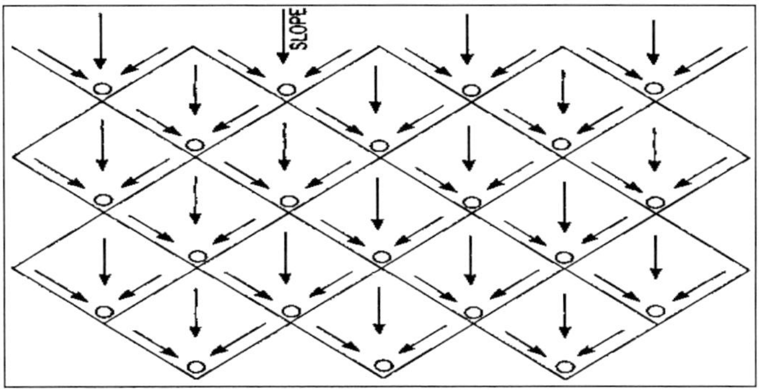

Figure 1. Negarim microcatchments - field layout

Negarim microcatchments are mainly used for tree growing in arid and semi-arid areas. Each microcatchment consists of a catchment area and an infiltration pit (cultivated area). The shape of each unit is normally square, but the appearance from above is of a network of diamond shapes with infiltration pits in the lowest corners (Figure 1).

While Negarim microcatchments are well suited for hand construction, they cannot easily be mechanised. Once the trees are planted, it is not possible to operate and cultivate with machines between the tree lines.

The area of each unit is either determined on the basis of a calculation of the plant (tree) water requirement or, more usually, an estimate of this.

Size of microcatchments (per unit) normally range between 10 m^2 and 100 m^2 depending on the specie of tree to be planted but larger sizes are also feasible, particularly when more than one tree will be grown within one unit.

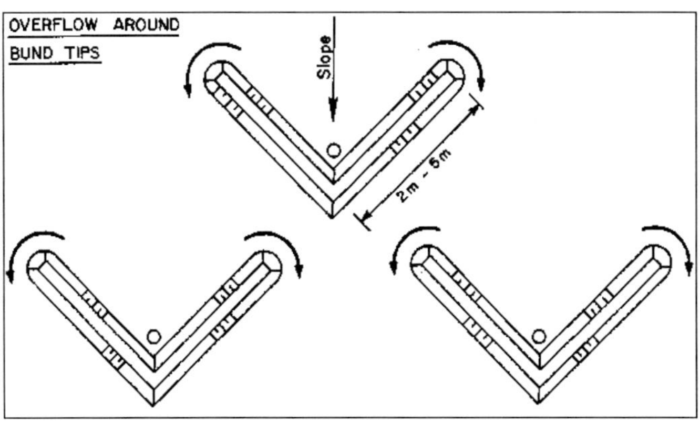

Figure 2."V"-shaped microcatchments

Bund Design

The bund height is primarily dependent on the prevailing ground slope and the selected size of the micro-catchment. It is recommended to construct bunds with a height of at least 25 cm in order to avoid the risk of over-topping and subsequent damage. Where the ground slope exceeds 2.0%, the bund height near the infiltration pit must be increased. The top of the bund should be at least 25 cm wide and side slopes should be at least in the range of 1:1 in order to reduce soil erosion during rainstorms. Whenever possible, the bunds should be provided with a grass cover since this is the best protection against erosion.

A common variation is to build microcatchments as single, open-ended structures in "V" or semi-circular shape (figure 2). The advantage is that surplus water can flow around the tips of the bunds, however, the storage capacity is less than that of a closed system. These types of bunds are particularly useful on broken terrain, and for small numbers of trees around homesteads.

Construction

Step one: The first step is to find a contour line. This can be done by using a line level or a water tube level. Since natural contours are often not smooth, it will be necessary to even out the contours so that finally a straight line is obtained. The first line, at the top of the block is marked (Figure 3). If the topography is very uneven, separate smaller blocks of microcatchments should be considered.

Step Two: By means of a tape measure, the tips of the bunds are now marked along the "straightened contour". The first line will be open-ended. The distance between the tips (a-b) depends on the selected catchment size.

Step Three: A piece of string as long as the side length of the catchment (5 m for a 5 m ´5 m microcatchment) is held at one tip (a) and a second string of the same length at the other tip (b). They will exactly meet at the apex (c). The apex is now marked with a peg and the catchment sides (a-c) and (b-c) marked on the ground alongside the strings with a hoe. This procedure will be repeated until all bund alignments in the first row have been determined.

Step Four: The next row of microcatchments can now be staked out. The apexes of the bunds of the upper row will be the tips for the second row and the corresponding apexed will be found according to Step 3. When the second row of microcatchments has been marked, repeat the same procedure for the third row, etc. The final result will be a block of diamond-shaped microcatchments, with a first row which is open at the upslope end.

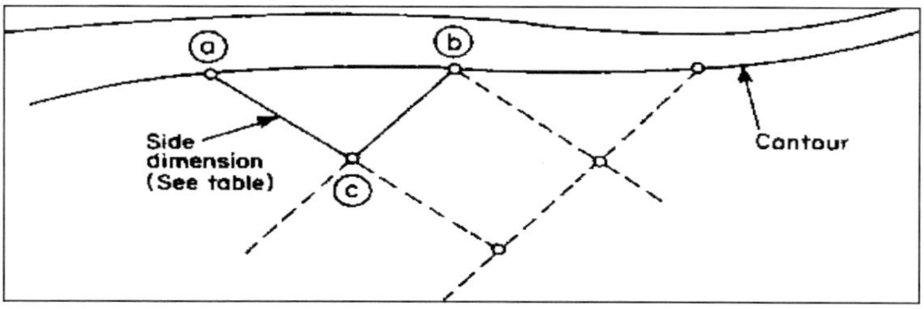

Figure 3. Negarim microcatchment: layout technique

Step Five: The size of the infiltration put is staked out and the pit is excavated—leaving a small step towards the back on which the seedling will be planted (Figure 4).

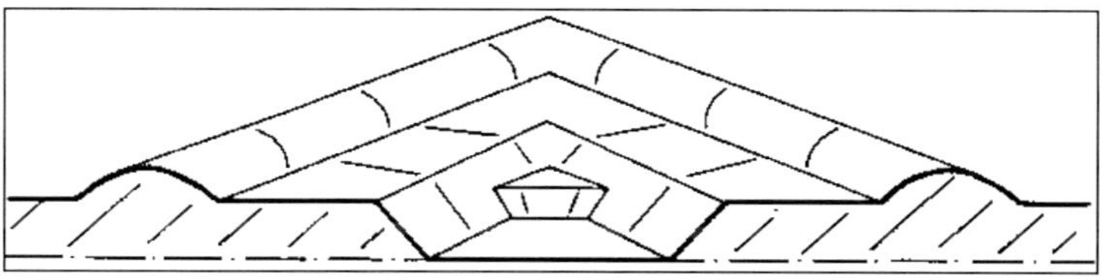

Figure 4. Infiltration pit with planting step

Step Six: Before constructing the bunds, the area within the microcatchments should be cleared of all vegetation. The bunds should then be constructed in two layers. The excavated material from the pit is used to form the bund. The bunds should be compacted during construction. Before compaction, the soil should be wetted wherever possible. Compaction may be done by foot or with a barrel filled with sand or water. To ensure a uniform height of the bund, a string should be fixed at the beginning and the end of each bund alignment and be adjusted above ground according to the selected bund height.

Step Seven: A diversion ditch should be provided above the block of microcatchments if there is a risk of damage by runoff from upslope of the block. The diversion ditch should be aligned in a 0.25% slope and in most cases a depth of 50 cm and a width of 1.0-1.5 m will be sufficient. The soil is deposited downslope. The diversion ditch should be constructed first to prevent damage in case a rainstorm occurs during construction of the microcatchments.

Maintenance will be required for repair of damages to bunds, which may occur if storms are heavy soon after construction when the bunds are not yet fully consolidated. The site should be inspected after each significant rainfall as breakages can have a "domino" effect if left unrepaired.

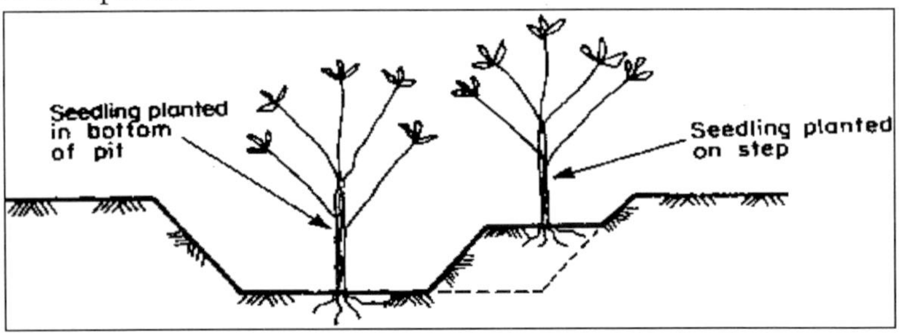

Figure 5. Planting site for seedling

Tree seedlings of at least 30 cm height should be planted immediately after the first rain of the season. It is recommended that two seedlings are planted in each microcatchment—one in the bottom of the pit (which would survive even in a dry year) and one on a step at the back of the pit. If both plants survive, the weaker can be removed after the beginning of the second season. For some species, seeds can be planted directly. This eliminates the cost of a nursery.

Manure or compost should be applied to the planting pit to improve fertility and water-holding capacity. If grasses and herbs are allowed to develop in the catchment area, the runoff will be reduced to some extent, however, the fodder obtained gives a rapid return to the investment in construction. Regular weeding is necessary in the vicinity of the planting pit.

Socio-economic Benefits

Negarim microcatchments have been developed in Israel for the production of fruit trees, but even there the returns on investment are not always positive. It is not a cheap technique, bearing in mind that one person-day is required to build (on average) two units, and costs per unit rise considerably as the microcatchment size increases. It is essential that the costs are balanced against the potential benefits. In the case of multipurpose trees in arid/semi-arid areas, for several years the main benefit will be the soil conservation effect and grass for fodder until the trees become productive. Negarim microcatchments are appropriate both in village afforestation blocks, or around homesteads where a few open-ended "V" shaped microcatchments provide shade or support amenity trees.

Growth and fruit production were significantly improved by microcatchment techniques. On 0.5% slopes a 72 m catchment area per tree was sufficient to obtain excellent fruit yields. On a 5% slope the size could be reduced to 54 m^2 because of the increased runoff. Runoff coefficients increased over years due to a soil crust formed over the microcatchment surface.

Contour Bunds

Contour bunds for trees are a simplified form of microcatchments. Construction can be mechanised and the technique is therefore suitable for implementation on a larger scale. As its name indicates, the bunds follow the contour, at close spacing, and by provision of small earth ties the system is divided into individual microcatchments. Whether mechanised or not, this system is more economical than Negarim microcatchment, particularly for large scale implementation on even land—since less earth has to be moved.

A second advantage of contour bunds is their suitability to the cultivation of crops or fodder between the bunds. As with other forms of microcatchment water harvesting techniques, the yield of runoff is high, and when designed correctly, there is no loss of runoff out of the system. Contour bunding for tree planting is not yet as common as Negarim microcatchments. Examples of its application come from Baringo District, Kenya.

Figure 6. Contour bunds for trees

The bunds are formed with soil excavated from an adjacent parallel furrow on their upslope side. Small earth ties perpendicular to the bund on the upslope side subdivide the system into microcatchments. Infiltration pits are excavated in the junction between ties and bunds. A diversion ditch protects the system where necessary.

The size of microcatchment per tree is estimated in the same way as for Negarim microcatchments. However, the system is more flexible, because the microcatchment size can be easily altered by adding or removing cross-ties within the fixed spacing of the bunds. Common sizes of microcatchments are around 10 -50 m^2 for each tree.

Bund Design

Bund heights vary, but are in the order of 20 - 40 cm depending on the prevailing slope. As bunds are often made by machine the actual shape of the bund depends on the type of machine; whether for example a disc plough or a motor grader is used. It is recommended that the bund should not be less than 25cm in height. Base width must be at least 75 cm. The configuration of the furrow upslope of the bund depends on the method of construction.

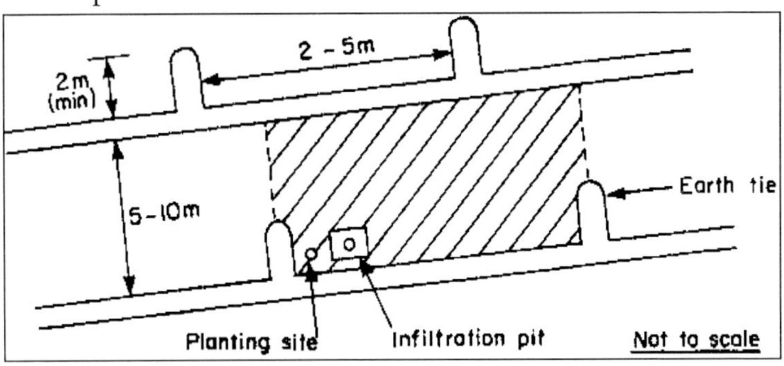

Figure 7. Microcatchment unit

Bunds should be spaced at either 5m or 10m apart. Cross-ties should be at least 2 metres long at a spacing of 2 to 10 metres. The exact size of each microcatchment is thus defined. It is recommended to provide 10 m spacing between the bunds on slopes of up to 0.5% and 5 m on steeper slopes. A common size of microcatchment for multipurpose trees is $25m^2$. This corresponds to 10 metre bund spacing with ties at 2.5m spacing or 5 metre bund-spacing with ties at 5 m spacing. Excavated soil from the infiltration pit is used to form the ties. The pit is excavated in the junction of the bund and the cross-tie. A pit size of 80 cm ´ 80 cm and 40 cm deep is usually sufficient.

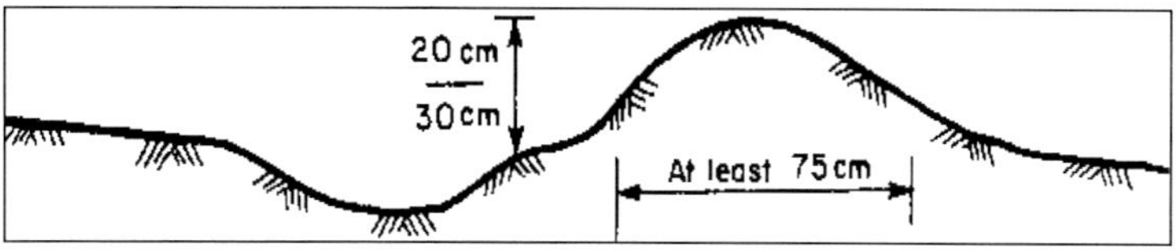

Figure 8. Bund dimensions

Table 1 gives quantities of earthworks required for various layouts of contour bunds. The bund height assumed is 25 cm with 75 cm base width.

Table 1. Quantities of earthworks

Size Unit Microcatchment			Volume Earthworks per Unit	No. Units per ha	Earthworks m^3/ha
Bund spac- ing	Tie spac ing	Area (m^2)	(m^3)		
5m	2m	10	0.5	1000	500
5m	5m	25	0.9	400	360
5m	10m	50	1.5	200	300
10m	2.5m	25	0.6	400	240
10m	5m	50	0.9	200	180

Construction

Step One: The contour is determined by means of a simple surveying instrument, such as a line level or a water tube level. As contour bunds are implemented on even land, contours need only be staked out approximately every 50 metres. The real contour should be smoothed to a gentle curve. Bunds may become slightly wider apart at one end to accommodate any change in the contour.

Step Two: The alignment of each bund should be marked on the ground before construction starts. As recommended under "bund design" the bunds should be set at a spacing of 10 m for slopes up to 0.5% and 5 m for steeper slopes. When disc ploughs are used, a single disc (with the remaining one or two removed) forms an adequate bund. Where available, a reversible plough is preferred because furrows can be ploughed consecutively in both directions. Compaction of the bunds is recommended. When no machines are available, this should be done by foot or with a barrel filled with sand.

Step Three: The catchment size required for each seedling determines the spacing between cross-ties. For example where a 25

m^2 catchment area is required and bunds are 10 metres apart the cross-ties will be 2.5 metres apart. Cross-ties are made by hand.

An infiltration pit of 80 cm □ 80 cm and 40 cm deep is dug in the furrow above the bund. Water collected in the furrow will then drain into the pit and supply the adjacent seedling. The excavated material is sufficient to form a cross-tie of 2 m length, 75 cm base width and 25 cm height. The cross-tie extends upslope from the main bund at an angle of 90° to the main bund; at least 30 cm should be left between the cross-tie and the pit to allow sufficient space for planting the seedling.

Step Four: At each side of the block, a lateral bund of 25 - 30 cm height is built to prevent loss of runoff out of the system. The earth should be excavated from inside the system and the contour bunds must be joined up with the lateral bund.

Step Five: A diversion ditch should be provided above the scheme if there is a risk of damage by runoff from outside the block. The diversion ditch is aligned on a 0.25% slope and a common dimension is 50 cm deep and 1 - 1.5 m wide, with the soil piled downslope. The diversion ditch should be constructed before the contour bunds are built to prevent damage if rainstorms occur during construction.

As with Negarim microcatchments, maintenance will in most cases be limited to repair of damage to bunds early in the first season. It is essential that any breaches -which are unlikely unless the scheme crosses existing rills—are repaired immediately and the repaired section compacted. Damage is frequently caused if animals invade the plots. Grass should be allowed to develop on the bunds, thus assisting consolidation with their roots.

The majority of the husbandry factors noted under Negarim microcatchments also apply to this system: there are, however, certain differences. Tree seedlings, of at least 30 cm height, should be planted immediately after the first runoff has been harvested. The seedlings are planted in the space between the infiltration pit and the cross-tie. It is advisable to plant an extra seedling in the bottom of the pit for the eventuality of a very dry year. Manure or compost can be added to the planting pit to improve fertility and water holding capacity. One important advantage of contour bunds for tree establishment is that oxen or mechanised cultivation can take place between the bunds, allowing crops or fodder to be produced before the trees become productive. However, this has the disadvantage of reducing the amount of runoff reaching the trees.

Socio-economic Benefits

Contour bunds for trees are mainly made by machine; costs of bund construction can be relatively low and implementation fast, especially where plots are large and even and the kind of mechanisation well adapted. However, as with all mechanisation in areas with limited resources, there is a question mark about future sustainability. Experience has shown that very often the machines come abruptly to a halt when the project itself ends. Another aspect that must be addressed is the management after the scheme has been established (which is usually done under the auspices of a development project). This is an issue which has to be seriously considered during the planning phase. Management of a large afforestation block by the local community is in most cases a new challenge and failure or success will depend on acceptance of the technique by the rural population.

Semi-circular Bunds

Semi-circular bunds are earth embankments in the shape of a semi-circle with the tips of the bunds on the contour. Semi-circular bunds, of varying dimensions, are used mainly for rangeland rehabilitation or fodder production. This technique is also useful for growing trees and shrubs and, in some cases, has been used for growing crops. Depending on the location, and the chosen catchment: cultivated area ratio, it may be a short slope or long slope catchment technique. The examples described here are short slope catchment systems.

Semi-circular bunds, are recommended as a quick and easy method of improving rangelands in semi-arid areas. Semi-circular bunds are more efficient in terms of impounded area to bund volume than other equivalent structures—such as trapezoidal bunds for example. Surprisingly, this technique has never been used traditionally.

The main limitation of semi-circular bunds is that construction cannot easily be mechanised. Since semi-circular bunds can be designed to a variety of dimensions. Design "a" comprises small structures, closely spaced. It is suitable for the relatively "wetter" semi-arid areas but requires low slopes and even terrain. Design "b", with larger and wider spaced bunds, is more suitable for drier areas, and does not need such even topography.

The two designs of semi-circular bunds considered here differ in the size of structure and in field layout. Design "a" has bunds with radii of 6 metres, and design "b" has bunds with radii of 20 metres. In both designs the semi-circular bunds are constructed in staggered lines with runoff producing catchments between structures. Design "a" is a short slope catchment technique, and is not designed to use runoff from outside the treated area, nor to accommodate overflow. Design "b" is also a short slope catchment system, but can accommodate limited runoff from an external source. Overflow occurs around the tips of the bund which are set on the contour.

C:CA ratios of up to 3:1 are generally recommended for water harvesting systems used for rangeland improvement and fodder - production. A detailed calculation is not required. The reasons for applying low ratios are that already adapted rangeland and fodder plants in semi-arid and arid areas need only a small amount of extra moisture to respond significantly with higher yields. Larger ratios would require bigger and more expensive structures, with a higher risk of breaching.

Design "a" as described here has a C:CA ratio of only 1.4:1, and does not require provision for overflow. Design "b" has a C:CA ratio of 3:1, and therefore provision for overflow around the tips of the bunds is recommended, though occurrence of overflow is usually rare. A larger C:CA ratio for design "b" is possible but it should not exceed 5:1.

Bund Design

Design "a":

This design, suitable for slopes of 1% or less, consists of a series of small semi-circular bunds with radii of 6 metres. Each bund has a constant cross section over the whole length of 19 m. The recommended bund height is 25 cm with side slopes of 1:1 which result in a base width of 75 cm at a selected top width of 25 cm.

The tips of each bund are set on the contour, and the distance between the tips of adjacent bunds in the same row is 3 metres. Bunds in the row below are staggered, thus allowing the collection of runoff from the area between the bunds above. The distance between the two rows, from the base of bunds in the first line to tips of bunds in the second, is 3 metres. At this spacing 70-75 bunds per hectare are required.

Design "b"

The radius of the semi-circle is 20 metres. The cross-section of the bund changes over its length. At the wing tip, the bund is only 10 cm high, but the height increases towards the middle of the base to 50 cm with side slopes of 3:1 (horizontal: vertical), and a top width of 10 cm. Corresponding base widths are 70 cm and 3.10 metres, respectively. As with design "a", the bunds must be arranged in a staggered configuration. Due to the larger dimensions of the bunds there are only 4 structures required per hectare.

The distance between the tips of two adjacent structures in one row is 10 m while 30 metres are recommended between the base of the upper structure and the tips of the lower one. Radii and distances between the structures can be increased or decreased according to the selected C:CA ratio. Design "b" is recommended on slopes up to 2%. For higher slopes, smaller radii are required. For example, on a slope of 4%, the radius should be reduced to 10 metres and the distance between two adjacent rows from 30 metres to 15 metres while the tips of two adjacent structures should be 5 m apart instead of 10 m. The number of structures required for one hectare would thus increase to 16 which maintains the C:CA ratio of 3:1.

Semi-circular bunds can be constructed in a variety of sizes, with a range of both radii and bund dimensions. Small radii are common when semi-circular bunds are used for tree growing and production of crops. A recommended radius for these smaller structures is 2 to 3 metres, with bunds of about 25 cm in height.

Construction

The layout for both designs is similar, only dimensions differ.

Step One: The first contour, at the top of the scheme, is staked out using a simple surveying instrument. This line need not be smooth.

Step Two: A tape measure is now used to mark the tips of the semi-circular bunds on the contour. For design "a", the tips of one structure are 12 metres apart (2 times the radius) and the distance to the next unit is 3 m. For design "b", the tips are 40 metres apart and the distance to the next structure is 10 m.

Step Three: The centre point between the tips of each semi-circular unit is marked. A piece of string as long as the selected radius is now fixed at the centre point by means of a peg. Holding the string tight at the other end, the alignment of the semi-circle is defined by swinging the end of the string from one tip to the other. The alignment can be marked by pegs or small stones (Figure 9).

Step Four: Staking out and construction of the semi-circular bunds in the second and all following rows will be carried out in the same way. It is important that the structures in each row are staggered in relation to structures in the row above. The centre points of the bunds, for example, in the second row should coincide with the middle of the gaps between bunds in the first row and so forth. It must be ensured that the space between bunds from one row to another is according to the chosen distance, that is, 3 m for design "a" and 30 m for design "b".

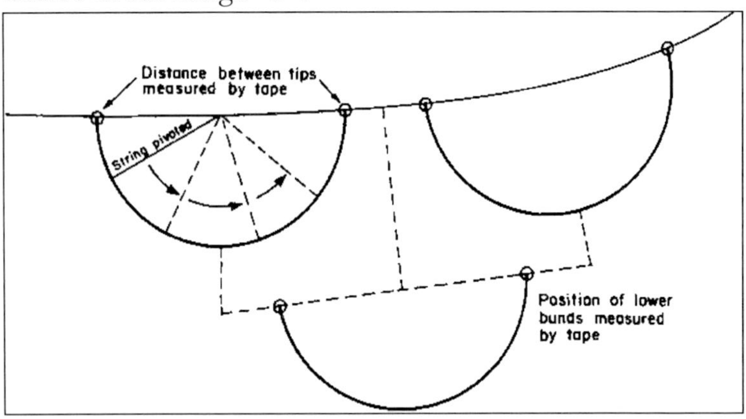

Figure 9. Layout technique

Step Five: After setting out, bund construction is started with excavation of a small trench inside the bund. Further excavation should always be from inside the bund, as evenly as possible. This will increase the storage capacity of the semi-circular bund. The bund should be constructed in layers of 10-15 cm with each layer being compacted and wetted first if possible.

Step Six: For design "b", it is recommended that the bund tips are protected with a layer of stones, as shown in Figure 10. This will ensure that the bund tips are more resistant to erosion when excess water discharges around them. A diversion ditch above the first row of structures may be necessary for design "a" to protect the first row of bunds against runoff coming from the catchment area above.

Where semi-circular bunds of design "a" are built in one block of several hectares it is advisable to provide one or more diversion ditches within the block as a safety factor. Diversion ditches should be 1 - 1.5 metre wide and 50 cm deep, with a gradient of 0.25%.

As with all earthen structures, the most critical period for semi-circular bunds is when rainstorms occur just after construction, since at this time the bunds are not yet fully consolidated. Any breakages must be repaired immediately. If damage occurs, it is recommended that a diversion ditch is provided if not already constructed. Semi-circular bunds which are used for fodder production normally need repairs of initial breaches only. This is because in the course of time, a dense network of the perennial grasses will protect the bunds against erosion and damage. The situation is different if animals have access into the bunded area and are allowed to graze. In this case, regular inspections and maintenance (repair) of bund damages will be necessary.

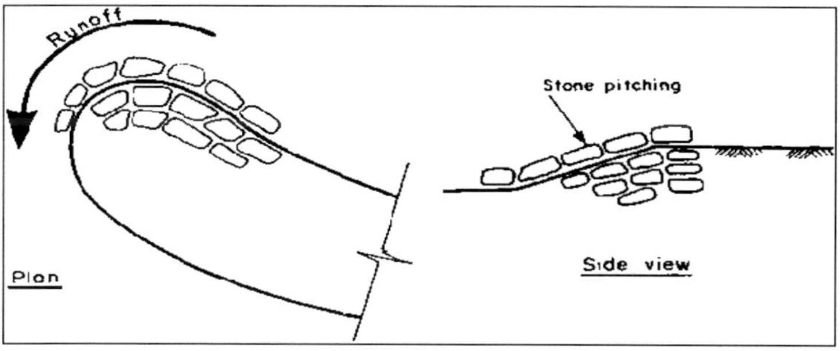

Figure 10. Protection of wingtips

Socio-economic Benefits

Water harvesting for range improvement and for fodder production will mainly be applied in areas where the majority of the inhabitants are agro-pastoralists—at least in the Sub-Saharan Africa context. In these areas, the concept of improving communally used rangeland is usually alien. Therefore, it may be difficult to motivate the population to invest voluntarily, in the time and effort required for implementing and maintaining such a water harvesting system. Even when this is possible it is equally important to introduce an appropriate and acceptable range management programme to avoid over-grazing and subsequent degradation of the range. Controlled grazing is also essential to maintain good quality rangeland, and the bunded area must be rested periodically for it to regenerate, so that natural reseeding can take place.

Contour Ridges for Crops

Contour ridges, sometimes called contour furrows or microwatersheds, are used for crop production. This is again a microcatchment technique. Ridges follow the contour at a spacing of usually 1 to 2 metres. Runoff is collected from the uncultivated strip between ridges and stored in a furrow just above the ridges. Crops are planted on both sides of the furrow. The system is simple to construct—by hand or by machine—and can be even less labour intensive than the conventional tilling of a plot.

The yield of runoff from the very short catchment lengths is extremely efficient and when designed and constructed correctly there should be no loss of runoff out of the system. Another advantage is an even crop growth due to the fact that each plant has approximately the same contributing catchment area. Contour ridges for crops are not yet a widespread technique. They are being tested for crop production on various projects in Africa.

Contour ridges are limited to areas with relatively high rainfall, as the amount of harvested runoff is comparatively small due to the small catchment area. The overall layout consists of parallel, or almost parallel, earth ridges approximately on the contour at a spacing of between one and two metres. Soil is excavated and placed downslope to form a ridge, and the excavated furrow above the ridge collects runoff from the catchment strip between ridges. Small earth ties in the furrow are provided every few metres to ensure an even storage of runoff. A diversion ditch may be necessary to protect the system against runoff from outside.

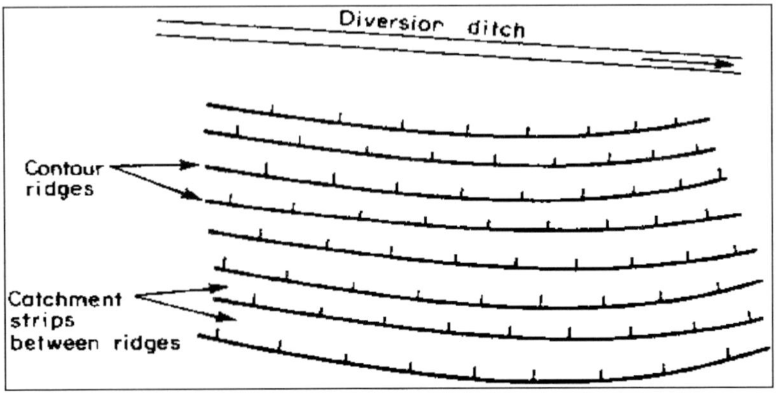

Figure 11. Contour ridges: field layout

Crops are planted within this zone, and use the runoff concentrated in the furrow. Thus for a typical distance of 1.5 m between ridges, the C:CA ratio is 2:1; that is a catchment strip of one metre and a cultivated strip of half a metre. A distance of 2 metres between ridges would give a 3:1 ratio. The C:CA ratio can be adjusted by increasing or decreasing the distance between the ridges. In practice a spacing of 1.5 - 2.0 metres between ridges (C:CA ratios of 2:1 and 3:1 respectively) is generally recommended for annual crops in semi-arid areas.

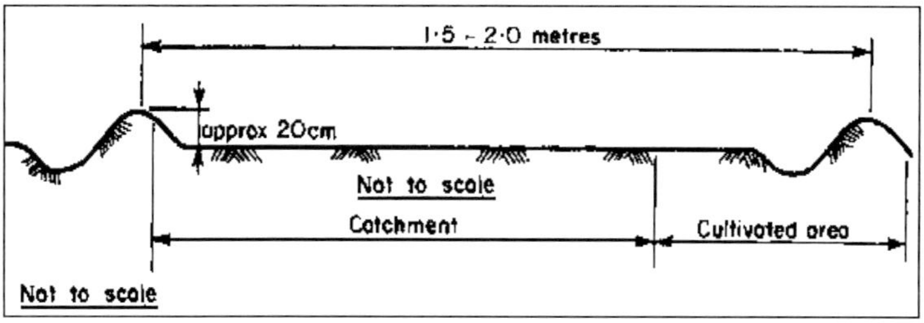

Figure 12. Contour ridge dimensions

Ridge Design

Ridges need only be as high as necessary to prevent overtopping by runoff. As the runoff is harvested only from a small strip between the ridges, a height of 15 -20 cm is sufficient. If bunds are spaced at more than 2 metres, the ridge height must be increased.

Table 2. Quantities of earthwork for contour Rides

Ridge spacing (m)	Ridge & Tie height (cm)	Earthworks per ha (m^3)
1.5	15	270
1.5	20	480
2.0	20	360

Quantities of earthwork for different contour ridge spacing and ridge heights are given below. It should be noted that the construction of the ridges already includes land preparation—no further cultivation is required. Where a diversion ditch is necessary, an additional 62.5 m3 for each 100 metre of length of ditch has to be added.

Construction

Step One: Contours are surveyed by a simple surveying instrument such as a water tube level or line level. The real contour should be smoothed to obtain a better alignment for agricultural operations.

Step Two: Contour keylines should be staked out every 10 or 15 metres. The alignment for the ridges is then marked in between the keylines according to selected spacing. On uneven terrain, the contours may come closer together at one point or widen at other points. It is necessary to stop lines where the contours converge or to add short extra lines in between where the contours diverge.

Step Three: The furrows are excavated usually by means of a hoe or are ploughed parallel to the marked alignments for the ridges. The excavated soil is placed downslope, next to the furrow, and the ridge is formed.

Step Four: Small cross-ties are built at intervals of about 5 metres dividing each furrow into a number of segments. The ties are 15-20 cm high and 50 - 75 cm long.

Step Five: A diversion ditch should be provided above the block of contour ridges if there is a risk of damage caused by runoff from outside the system. The diversion ditch should be 50 cm deep and 1-1.5 m wide, with a gradient of 0.25%. The excavated soil is placed downslope. The ditch should be constructed before the contour ridges are built to prevent damage from early rains.

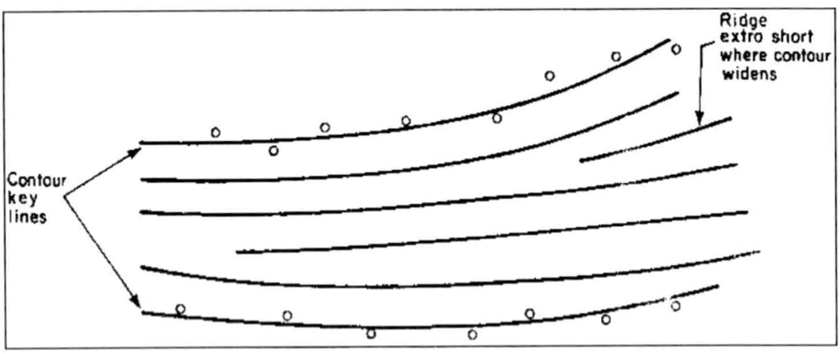

Figure 13. Contour ridges: layout technique

If contour ridges are correctly laid out and built, it is unlikely that there will be any overtopping and breaching. Nevertheless if breaches do occur, the ridges or ties must be repaired immediately. The uncultivated catchment area between the ridges should be kept free of vegetation to ensure that the optimum amount of runoff flows into the furrows. At the end of each season the ridges need to be rebuilt to their original height. After two or three seasons, depending on the fertility status of the soils, it may be necessary to move the ridges downslope by approximately a metre or more, which will result in a fresh supply of nutrients to the plants.

The main crop (usually a cereal) is seeded into the upslope side of the ridge between the top of the ridge and the furrow. At this point, the plants have a greater depth of top soil. An intercrop, usually a legume, can be planted in front of the furrow. It is recommended that the plant population of the cereal crop be reduced to approximately 65% of the standard for conventional rainfed cultivation. The reduced number of plants thus have more moisture available in years of low rainfall.

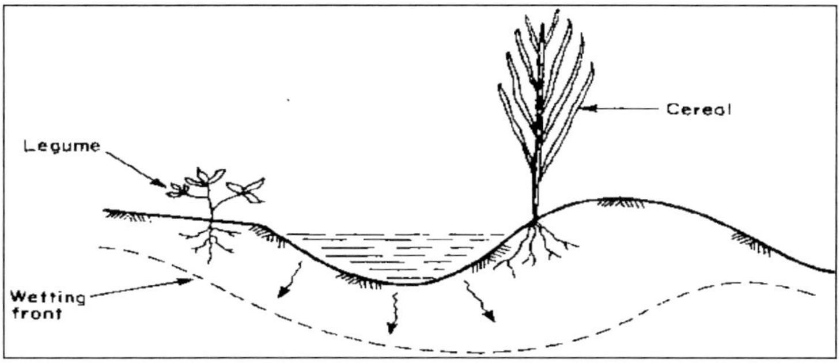

Figure 14. Planting configuration

Socio-economic Benifits

Since the contour ridge technique implies a new tillage and planting method compared with conventional cultivation, farmers may be initially reluctant to accept this technique. Demonstration and motivation are therefore very important. On the other hand, it is one of the simplest and cheapest methods of water harvesting. It can be implemented by the farmer using a hoe, at no or little extra cost. External support is limited to a minimum. Alternatively it can be mechanised and a variety of implements can be used. When used by a farmer on his own land, the system does not create any conflicts of interest between the implementor and the beneficiary.

Trapezoidal Bunds

Trapezoidal bunds are used to enclose larger areas (up to 1 ha) and to impound larger quantities of runoff which is harvested from an external or "long slope" catchment. The name is derived from the layout of the structure which has the form of a trapezoid -a base bund connected to two side bunds or wingwalls which extend upslope at an angle of usually 135. Crops are planted within the enclosed area. Overflow discharges around the tips of the wingwalls. The general layout, consisting of a base bund connected to wingwalls is a common traditional technique in parts of Africa. The concept is similar to the semi-circular bund technique: in this case, three sides of a plot are enclosed by bunds while the fourth (upslope) side is left open to allow runoff to enter the field. The simplicity of design and construction and the minimum maintenance required are the main advantages of this technique.

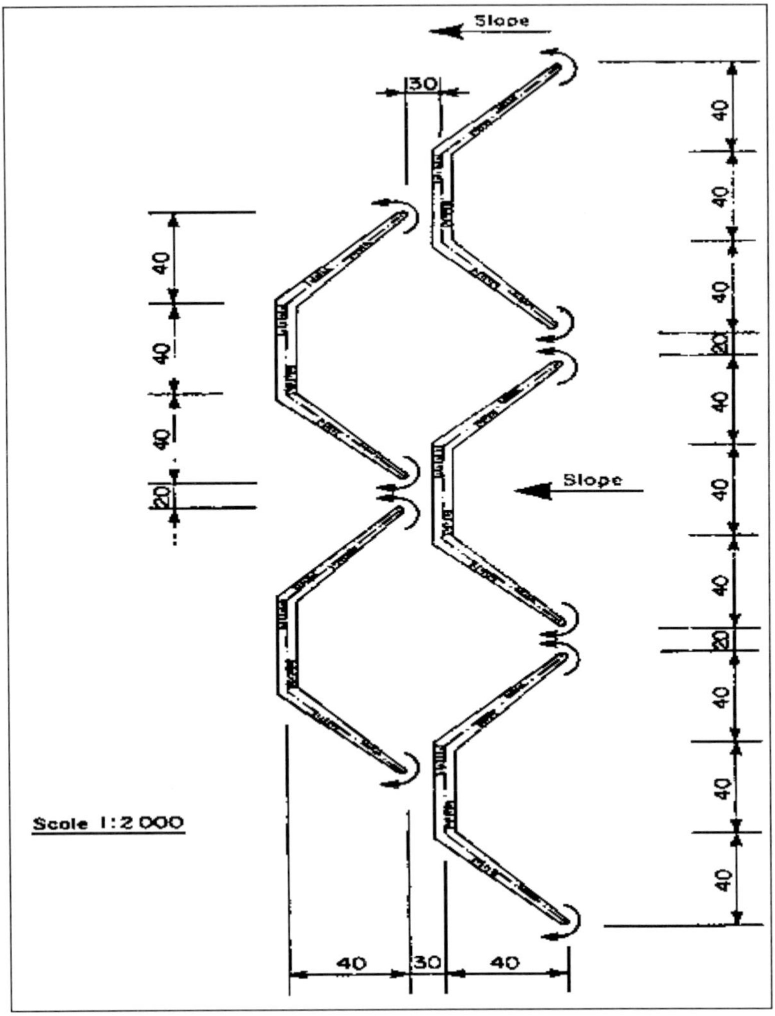

Figure 15. Trapezoidal bunds: field layout for 1% groundslope

In common with all water harvesting techniques which rely on external catchments, the C:CA ratio is based on seasonal rainfall reliability in a year of relatively low rainfall. In years of high rainfall, and particularly under storm conditions resulting in excessive inflow, damage can be caused to crops and to the bunds themselves. This is particularly the case for bunds on steeper slopes and for those with high C:CA ratios. This results in recommendation of a maximum C:CA ratio of 10:1, although ratios of up to 30:1 are sometimes used. Where the use of a large catchment is unavoidable a temporary diversion ditch is advisable to prevent excessive inflow of runoff. Conversely, in situations where the catchment is not of adequate size, interception ditches can be excavated to lead runoff from adjacent catchments to the bunds.

The configuration of the bunds is dependent upon the land slope, and is determined by the designed maximum flooded depth of 40 cm at the base bund. Traditional forms of water harvesting, similar to trapezoidal bunds, are found in the clay plains of Eastern Sudan and also in Somalia. In Sudan, the layout of the bunds is rectangular with the wingwalls extending upslope at right angles to the base bund. In Northwest Somalia a development project has constructed banana-shaped bunds by bulldozer.

Construction

Step One: When the site for the bund has been decided, the first thing to do is to establish the land slope using an Abney level or line level. Having established the ground slope, the tips of the wingwalls will be determined. Starting at the top of the field a peg is placed which will be the tip of one of the wingwalls.

Step Two: The dimensions for staking out the four main points of the bund—Point "a" can be established by measuring the distance "x" from Point 1 along the line joining points 1 and 2. Similarly, Point "b" is established by measuring the distance "x" from Point 2 along the line joining points 1 and 2. Points 3 and 4, which are the points of intersection of the base bund and the wingwalls, lie a distance □ downslope from points "a" and "b" respectively, measured at right angles to the line joining Point 1 and Point 2. The right angle can most easily be found by using a wooden right angle triangular template (sides: 100 cm, 60 cm and 80 cm). Points 3 and 4 are then pegged.

Step Three: The accuracy of the setting out can be checked by measuring the distances between Point 3 and Point 4, Point 3 and Point 1, and Point 2 and Point 4 which should be:

Point 3 - Point 4 = 40 m

Point 1 - Point 3 = Point 2 - Point 4

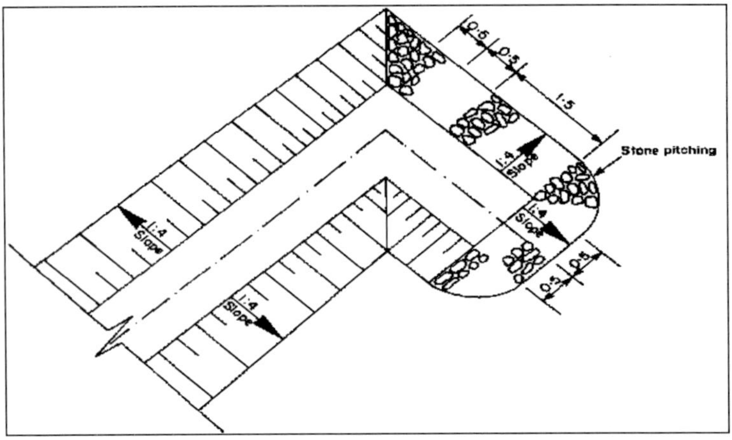

Figure 16. Trapezoidal bund: detail of tip

If, on checking, there is an error greater than 0.5 m in any of these three dimensions, the setting out procedure should be repeated.

Step Four: Having set the main points of the bunds it is necessary then to set out pegs or stones to mark the earthworks limits. Along the base bund this is done by marking parallel lines at a distance of 2.9 m from Line 3-4. For the wingbunds, the demarcation of the earthworks limits is slightly more complicated. At point 1 (2) perpendicular distances of 1.30 m either side of the wingbund centreline are measured and marked. At point 3 (4) distances of 2.90 m either side and perpendicular to the wingbunds centreline (line 1-3, 2-4) are measured and marked. It is then possible to peg out earthwork limits on both sides of the wingbunds centreline. Where more than one bund is required, the other bunds should be pegged out accordingly.

Step Five: Construction of a set of trapezoidal bunds must start with the row furthest upslope. Before commencing construction the soil within the foundation area of the bunds should be loosened to ensure good "mating" with the fill, The bund is constructed in two layers, each having a maximum thickness of 0.30 m. The thickness of the first layer will gradually taper off to zero as filling proceeds upslope along the wingbunds. Similarly, the thickness of the second layer will taper to 0.20 m at the tips- Each layer should be thoroughly compacted by rolling, ramming or stamping, and should be watered prior to compaction, where this is possible.

Excavation to provide the necessary fill should be taken from within the bunded area, where possible, to assist in levelling the area within the bund to promote even depth of flooding. Material for fill should not be excavated adjacent to the bunds on their downslope side, as this promotes gullying and bund failure.

Step Six: The tips of the bunds are only 20 cm high, and excess runoff drains around them. To prevent erosion of the tips they should be shaped with a small extension or "lip" to lead water away. This lip should be pitched with stones for extra resistance to erosion.

Step Seven: Where the catchment is large in relation to the bunded area, it is advisable to construct a diversion ditch to prevent excessive inflow to the bunds. This ditch is usually 50 cm deep and of 1.0 to 1.5 metres width, and is usually graded at 0.25%. Soil excavated from the ditch is used to construct an embankment on the downslope side, which also assists in diverting runoff from the bunds. During the early part of the season breaches can be made in this embankment at approximately 10 metre intervals and the material used to temporarily plug the ditch, thus permitting runoff to enter the trapezoidal bunds.

Step Eight: In situations where the catchment is not of adequate size, interception ditches can be made to lead runoff from adjacent catchments to the bunds. These are opposite in effect to diversion ditches but have similar sizes and design criteria.

If there are breaches in the bund, these must be repaired immediately, and the earth compacted thoroughly. Breaches are often caused by poor construction, or because the catchment area is producing damaging amounts of runoff - or both. It is advisable to construct a diversion ditch to protect the repaired bund. Holes burrowed by rodents can be another cause of breaching. These should be filled in whenever spotted. Allowing natural vegetation to grow on the bunds leads to improved consolidation by the plant roots. Repairs to the wing tips will frequently be needed when overflow has occurred. These should be built up, and extra stone pitching provided if required.

Trapezoidal bunds are normally used for production of annual crops in dry areas. The most common crops are cereals, and of these sorghum and bulrush (pearl) millet are by far the most usual. Sorghum is particularly appropriate for such systems because it is both drought tolerant and withstands temporary waterlogging. In the trapezoidal bund, water tends to be unevenly distributed because of the slope, and ponding often occurs near the base bund. Likewise the upper part may be relatively dry. Sorghum can tolerate both these situations.

Planting is carried out in the normal way, after ordinary cultivation of the soil within the bund. It is usual to plough parallel to the base bund, so that the small furrows formed by ploughing will locally accumulate some water. In the driest areas planting is sometimes delayed until a runoff event has saturated the soil within the bund, and germination/establishment is guaranteed. It is also possible to make use of out-of-season showers by planting a quick maturing legume, such as cowpea or tepary beans (Phaseolus acutifolius). Another useful technique is to plant curcurbits like gourds or watermelons on the bottom bund if water ponds deeply.

Socio-economic Benefits

It is difficult to generalize about the socio-economic factors concerning trapezoidal bunds, as different variations are found in different circumstances. As mentioned previously, there are examples of similar structures being used traditionally in Sudan -where they are often made by hand, without assistance from any agency and evidently perform well. On the other hand trapezoidal or similar bunds have been installed in other places under projects using food-for-work labour or even heavy machinery. When this has been done without any significant beneficiary commitment the bunds have been quickly abandoned. The amount of earthmoving necessary for trapezoidal bunds means that their construction usually requires organised labour or machinery and is beyond the scope of the individual farmer. However, where adequate motivation exists, there is considerable scope for the technique which has a traditional basis and does not require new farming skills.

Contour Stone Bunds

Contour stone bunds are used to slow down and filter runoff, thereby increasing infiltration and capturing sediment. The water and sediment harvested lead directly to improved crop performance. This technique is well suited to small scale application on farmer's fields and, given an adequate supply of stones, can be implemented quickly and cheaply. Making bunds—or merely lines—of stones is a traditional practice in parts of Sahelian West Africa, notably in Burkina Faso. Improved construction and alignment along the contour makes the technique considerably more effective. The great advantage of systems based on stone is that there is no need for spillways, where potentially damaging flows are concentrated. The filtering effect of the semi-permeable barrier along its full length gives a better spread of runoff than earth bunds are able to do. Furthermore, stone bunds require much less maintenance.

Stone bunding techniques for water harvesting (as opposed to stone bunding for hillside terracing, a much more widespread technique) is best developed in Yatenga Province of Burkina Faso. It has proved an effective technique, which is popular and quickly mastered by villagers.

Construction

Step One: The average slope of the field is determined by a simple surveying instrument such as a water level or a line level to decide on the spacing of the bunds. Each contour line is then set out and pegged individually. A horizontal spacing of approximately 20 metres apart is recommended for slopes of up to 1%, and 15 metres apart for 1 - 2% slopes. Because of variations in the slope, the lines may come closer together, or diverge at some points. The horizontal spacings recommended are the average distances apart. If labour is a limiting factor, farmers can start with a single bund at the bottom of their fields and work progressively upslope in seasons to come.

Step Two: After the exact contour is laid out, the line should be smoothed by moving individual pegs up or downslope. As a guideline, for ground slopes of up to 1.0 % pegs can be moved 2 metres upslope or downslope to create a smoother curve. Not only will a gentle curve be easier to follow while ploughing, but also the amount of stone used for construction will be reduced.

Step Three: A shallow trench is now formed along the smoothed contour. The trench is made by hand tools, or ploughed by oxen and then excavated by hand. The trench need only be 5 -10 cm deep and equal to the base width of the bund (35 - 40 cm). The excavated soil is placed upslope.

Step Four: Construction begins with large stones laid down at the base and the downslope side of the trench, and then smaller stones laid in front and on top of this "anchor" line. Small stones should be used to plug gaps between the larger ones. Where possible, a line of small or gravelly stones should run along the upslope face of the bund to create a fine filter. The key to a successful stone bund is to eliminate any large gaps between stones. In some areas it will be necessary to break large stones to produce the correct sizes of material.

During heavy runoff events stone bunds may be overtopped and some stones dislodged. These should be replaced. A more common requirement is to plug any small gaps with small stones or gravel where runoff forms a tunnel through. Eventually stone bunds silt-up, and their water harvesting efficiency is lost. It normally takes 3 seasons or more to happen, and occurs more rapidly where bunds are wider apart, and on steeper slopes. Bunds should be built up in these circumstances with less tightly packed stones, to reduce siltation, while maintaining the effect of slowing runoff. Alternatively grasses can be planted alongside the bund. Andropogon guyanus is the best grass for this purpose in West Africa. It can be seeded, and the mature grass is used for weaving into mats. The grass supplements the stone bund and effectively increases its height.

Socio-economic Benefits

On-farm stone bunding for crop production is quickly appreciated and adopted by farmers. The techniques involved, including simple surveying, can be easily learned. The amount of labour required is reasonable, and where groups are organised to work in turn on individual member's farms, fields can be transformed in a single day. The benefits of stone bunding are often clearly seen already in the first season—and this helps to make the system popular. Nevertheless there are some problems which must be faced. Relatively rich farmers can make use of wage labour to treat their fields, and poorer farmers may lag behind.

Permeable Rock Dams

A permeable rock dam is a long, low structure, made from loose stone (occasionally some gabion baskets may be used) across a valley floor. The central part of the dam is perpendicular to the watercourse, while the extensions of the wall to either side curve back down the valleys approximately following the contour. The idea is that the runoff which concentrates in the centre of the valley, creating a gully, will be spread across the whole valley floor, thus making conditions more favourable for plant growth. Excess water filters through the dam, or overtops during peak flows. Gradually the dam silts up with fertile deposits. Usually a series of dams is built along the same valley floor, giving stability to the valley system as a whole.

The calculation of the C:CA ratio is not necessary as the catchment area and the extent of the cultivated land are predetermined. However, the catchment characteristics will influence the size of structure and whether a spillway is required or not.

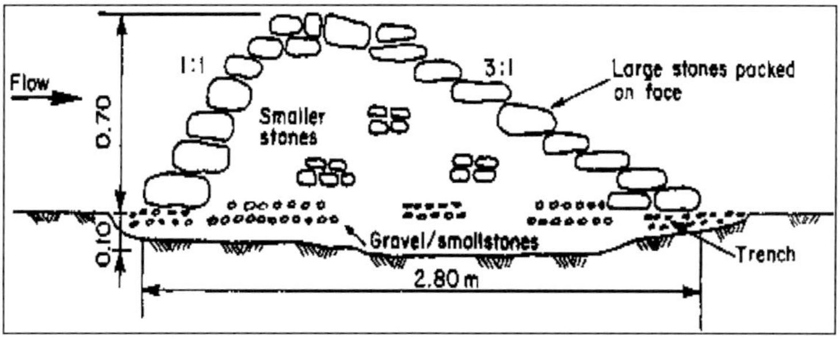

Figure 17. Dam dimensions

The main part of the dam wall is usually about 70 cm high although some are as low as 50 cm. However, the central portion of the dam including the spillway (if required) may reach a maximum height of 2 m above the gully floor. The dam wall or "spreader" can extend up to 1000 metres across the widest valley beds, but the lengths normally range from 50 to 300 metres. The amount of stone used in the largest structures can be up to 2000 tons. The dam wall is made from loose stone, carefully positioned, with larger boulders forming the "framework" and smaller stones packed in the middle like a "sandwich". The sideslopes are usually 3:1 or 2:1 (horizontal: vertical) on the downstream side, and 1:1 or 1:2 on the upstream side. With shallower side slopes, the structure is more stable, but more expensive. For all soil types it is recommended to set the dam wall in an excavated trench of about 10 cm depth to prevent undermining by runoff waters. In erodible soils, it is advisable to place a layer of gravel, or at least smaller stones, in the trench.

The quantity of stone, and the labour requirement for collection, transportation and construction depends on a number of factors and vary widely. Transport of stones by lorries from the collection site to the fields in the valley is the normal method. Considerable labour may be required to collect, and sometimes break, stone. Labour requirements, based on field estimates, are in the range of 0.5 cubic metres of stone per person/day—excluding transport. Where permeable rock dams are constructed in wide, relatively flat valley floors, they are sometimes made straight across - in contrast to the usual design where the spreader bunds arch back from the centre to follow the contour. With straight dams, the height of the wall decreases from the centre towards the sides of the valley to maintain a level crest.

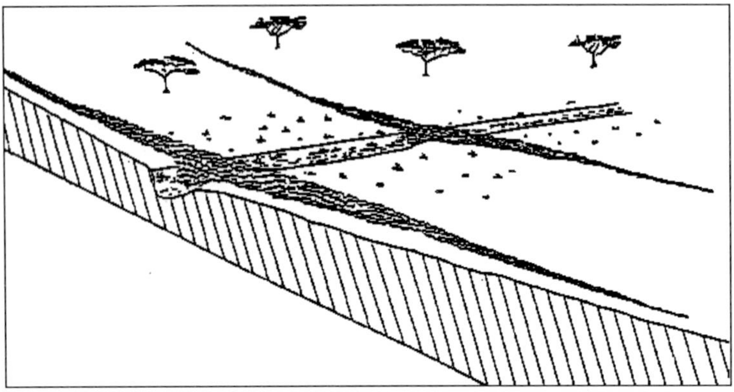

Figure 18. Permeable rock dams: alternative layout

Permeable rock dams are similar in many respects to the "terraced wadis" traditionally used in North Africa and the Middle East. However, the terraced wadi system is used in more arid regions, across clearly defined watercourses. Cross-wadi walls of stone retain runoff to depths of up to 50 cm, with the excess flowing over spillways into successive terraces below. Crops and fruit trees utilise the residual moisture.

Construction

Step One: Site selection depends both on the beneficiaries and the technicians. Theoretically it is best to start at the top of the valley, though this may not always be the people's priority. After site identification it is necessary to determine whether the structure needs a defined spillway: as a rule of thumb no spillway is required if the gully is less than one metre deep. For greater depths, a spillway is recommended. Gullies of over two metres depth pose special problems and should be only tackled with caution. It is important not to build a permeable rock dam immediately above a gully head, as there is the risk that the dam will fall into the gully if continued erosion causes the gully head to cut back.

Step Two: Where a spillway is required, this should be built first. Gabions are best for spillways, as loose stone is easily destabilised by heavy floods. The following should be noted:

a. A foundation of small stones, set in a trench, is required.
b. An apron of large rocks is needed to break the erosive force of the overflow.
c. The downstream banks of the watercourse should be protected by stone pitching to prevent enlargement of the gully.

Step Three: The alignment of the main dam walls can be marked out, starting at the centre of the valley (where there may/may not be a spillway). This alignment is ideally along the contour, or as close to the contour as possible. Thus the extension arms sweep backwards in an arc like the contours of a valley on a map. The arms end when they turn parallel to the watercourse. The contour can be laid out simply using a water tube or line level.

Step Four: A typical cross section is recommended for general use. This is of 280 cm base width, 70 cm height and side slopes of 1:1 upstream and 3:1 downstream. Larger cross sections may be required dependent on catchment characteristics. The first action after aligning the extension arms of the dam is to dig a trench at least 10 cm deep and 280 cm wide (according to the base width of the bund). The earth should be deposited upslope and the trench filled with gravel or small stones.

Step Five: The skill of construction is in the use of large stones (preferably of 30 cm diameter or more) for the casing of the wall. This should be built up gradually following the required sideslope, and the centre packed with smaller stones. The whole length of the bund should be built simultaneously, in layers. This layered approach reduces the risk of damage by floods during construction.

Earth should not be mixed with the stone because it may be washed out and thus destabilise the structure. It is particularly important to pack the small stones well at the lower levels to increase the rate of siltation. The structure is finished off with a cap of large stones. It should be possible to walk on the structure without any stones falling off. The dam wall should be level throughout its length, which can be checked by the use of a water tube or line level.

Step Six: If a series of permeable rock dams is to be built, an appropriate vertical interval (VI) should be selected. Technically speaking it is correct to:

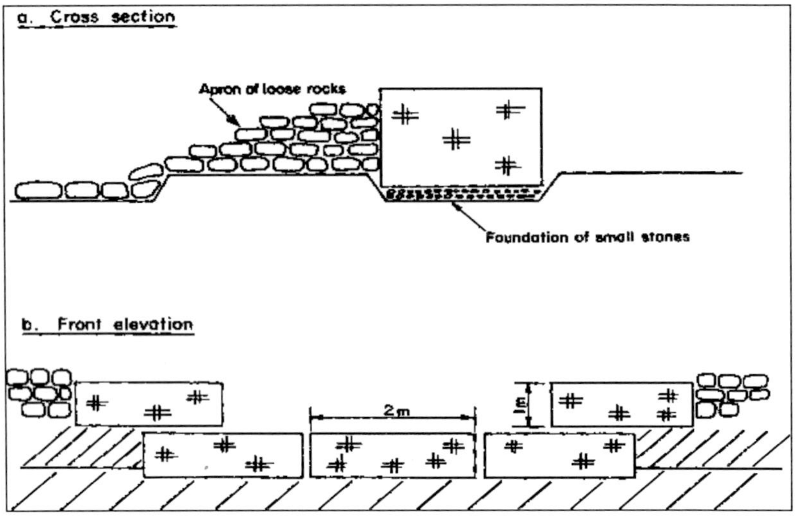

Figure 19. Gabion spillway

a. start at the top of the valley and work down;

b. use a VI equal to the height of the structure - so that the top of one structure is at the same level as the base of the one above it (Figure 20).

Therefore for dams of 70 cm height, the VI should theoretically be 70 cm. However in practice this may not be practicable due to the amount of stone and labour involved. As a compromise, a V.I. of 100 cm might be more realistic. Even wider spacing could be adopted, and the "missing" structures "filled in" afterwards. The vertical interval can be determined most easily by the use of a line-level. The horizontal spacing between adjacent dams can be determined from the selected VI and the prevailing land slope according to the formula:

$$HI = (VI \times 100)/\%slope$$

where:

HI = horizontal interval (m)

VI = vertical interval (m)

% slope = land gradient expressed as a percentage. For example, for a VI of 0.7 m and a 1% land slope,

$$HI = (0.7 \times 100)/1 = 70 \text{ metres}$$

For a VI of 0.7 m and a 2% land slope,

$$HI - (0.7 \times 100)/2 = 35 \text{ metres}$$

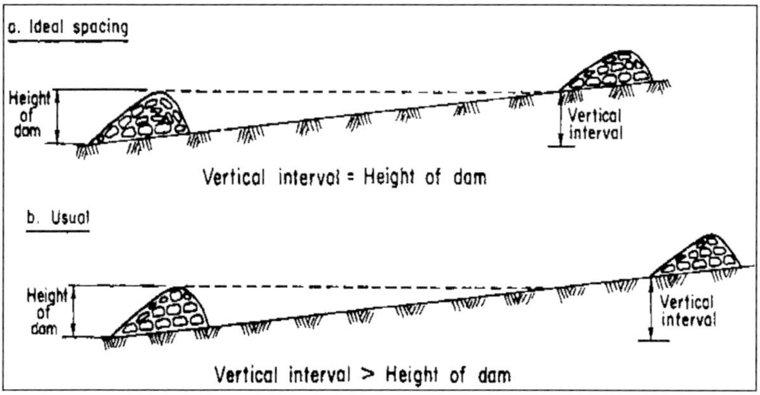

Figure 20. Spacing of rock dams

Permeable rock dams improve conditions for plant growth by spreading water, where moisture availability is a limiting factor. In addition, sediment, which will build up behind the bund over the seasons, is rich in nutrients, and this will further improve the crop growth. This technique is used exclusively for annual crops. In the sandier soils, which do not retain moisture for long, the most common crops are millet and groundnuts. As the soils become heavier, the crops change to sorghum and maize. Where soils are heavy and impermeable, waterlogging would affect most crops, and therefore rice is grown in these zones. Within one series of permeable rock dams, several species of crop may be grown, reflecting the variations in soil and drainage conditions.

Socio-economic Benefits

The implementation of permeable rock dams raises several important socio-economic issues. Many of these are rather specific to this technique. This is because permeable rock dams are characterised by:

a) large quantities of stone needed;
b) outside assistance often necessary for transport of stone;
c) limited number of direct beneficiaries;
d) siting is often determined by the people rather than the technicians.

As the structures cannot be made by individual farmers, it is necessary to cooperate in construction. It would be ideal if a village committee can be formed to co-ordinate efforts and discuss the situation of priority sites and beneficiaries. It is unrealistic to expect implementation of such a programme without outside help for transport of stones, which should be provided free of charge to the beneficiaries. Long-term sustainability and replicability of the form of development would best be promoted if beneficiaries could establish revolving funds for the hire or purchase of transport.

References

Ben-asher, J., "A review of water harvesting in Israel", (Draft) *Working Paper for World Bank's Sub-Saharan Water Harvesting Study*, 1988.

Billy, B., "Water harvesting for dryland and floodwater farming in the Navajo Indian reservation", In: Dutt, G.R. et al (eds), *Rainfall collection for agriculture in arid and semi-arid regions: 3-8*, Commonwealth Agricultural Bureaux, UK, 1981.

Boers, T.M. and Ben-Asher, J., "Review of rainwater harvesting", *In: Agricultural Water Management* 5: 145-158, 1982.

Bruins, H.J. Evanari, M. and Nessler, U., "Rainwater harvesting agriculture for food production in arid zones: the challenge of the African famine", *In: Applied Geography* 6 (I): 13-33, 1986.

Critchley, W.R.S., "Runoff harvesting for crop production: experience in Kitui District; 1984-1986", *Paper presented to the Third National Soil and Water Conservation Workshop*, Nairobi, Kenya, 1986.

Dutt, G.R., Hutchinson, C.F. and Garduno, M.A. eds., "Rainfall collection for agriculture in arid and semi-arid regions", *Proceedings of a workshop*, University of Arizona and Chapino Postgraduate College. Commonwealth Agric. Bureaux, UK, 1981.

Evanari, M., Shanan, L. and Tadmor, N.H., *Runoff farming in the Negev desert of Israel: 2nd Progress Report Avdat and Shivta farm project*, 1962/65. Rehovot, Israel, 1964.

CHAPTER 5

Rainwater Harvesting

Rainwater harvesting is a technology used for collecting and storing rainwater for human use from rooftops, land surfaces or rock catchments using simple techniques such as jars and pots as well as engineered techniques. Rainwater harvesting has been practiced for more than 4,000 years, owing to the temporal and spatial variability of rainfall. It is an important water source in many areas with significant rainfall but lacking any kind of conventional, centralised supply system. It is also a good option in areas where good quality fresh surface water or groundwater is lacking. The application of appropriate rainwater harvesting technology is important for the utilisation of rainwater as a water resource.

Global population has more than doubled since 1950 and reached six billion in 1999. The most recent population forecasts from the United Nations indicate that, under a medium-fertility scenario, global population is likely to peak at about 8.9 billion in 2050. Given that many natural resources (such as water, soil, forests and fish stocks) are already being exploited beyond their limits in some regions, significant effort will be required to meet the needs of an additional three billion people in the next 50 years. In parallel with these changes, there have been profound demographic shifts as people continue to migrate from rural to urban areas in search of work and new opportunities.

Since 1950, the number of people living in urban areas has jumped from 750 million to more than 2.5 billion people. Currently, some 61 million people are added to cities each year through rural to urban migration, natural increase within cities, and the transformation of villages into urban areas. Urbanisation creates new needs and aspirations, as people work, live, move and socialise in different ways, and require different products and services. Urban environmental impacts and demands are also different. By 2025, the total urban population is projected to double to more than five billion, and 90 per cent of this increase is expected to occur in developing countries.

Rapid population growth, combined with industrialisation, urbanisation, agricultural intensification and water-intensive lifestyles is resulting in a global water crisis. About 20 per cent of the population currently lacks access to safe drinking water, while 50 per cent lacks access to a safe sanitation system. Falling water tables are widespread and cause serious problems, both because they lead to water shortages and, in coastal areas, to salt intrusion. Both contamination of drinking water and nitrate and heavy metal pollution of rivers, lakes and reservoirs are common problems throughout the world. The world supply of freshwater cannot be increased. More and more people are becoming dependent on limited supplies of freshwater that are becoming more polluted. Water security, like food security, is becoming a major national and regional priority in many areas of the world. By the year 2025, two thirds of the world population may be subject to water scarcity.

Historical Development

Rainwater harvesting and utilisation systems have been used since ancient times and evidence of roof catchment systems date back to early Roman times. Roman villas and even whole cities were designed to take advantage of rainwater as the principal water source for drinking and domestic purposes since at least 2000 B.C. In the Negev desert in Israel, tanks for storing runoff from hillsides for both domestic and agricultural purposes have allowed habitation and cultivation in areas with as little as 100mm of rain per year. The earliest known evidence of the use of the technology in Africa comes from northern Egypt, where tanks ranging from 200-2000m^3 have been used for at least 2000 years – many are still operational today.

The technology also has a long history in Asia, where rainwater collection practices have been traced back almost 2000 years in Thailand. The small-scale collection of rainwater from the eaves of roofs or via simple gutters into traditional jars and pots has been practiced in Africa and Asia for thousands of years. In many remote rural areas, this is still the method used today. The world's largest rainwater tank is probably the Yerebatan Sarayi in Istanbul, Turkey. This was constructed during the rule of Caesar Justinian (A.D. 527-565). It measures 140m by 70m and has a capacity of 80,000 cubic metres.

Advantages

Rainwater harvesting systems can provide water at or near the point where water is needed or used. The systems can be both owner and utility operated and managed. Rainwater collected using existing structures (i.e., rooftops, parking lots, playgrounds, parks, ponds, flood plains, etc.), has few negative environmental impacts compared to other technologies for water resources development. Rainwater is relatively clean and the quality is usually acceptable for many purposes with little or even no treatment. The physical and chemical properties of rainwater are usually superior to sources of groundwater that may have been subjected to contamination. Some other advantages of rainwater harvesting include:

i) Rainwater harvesting can co-exist with and provide a good supplement to other water sources and utility systems, thus relieving pressure on other water sources.

ii) Rainwater harvesting provides a water supply buffer for use in times of emergency or breakdown of the public water supply systems, particularly during natural disasters.

iii) Rainwater harvesting can reduce storm drainage load and flooding in city streets.

iv) Users of rainwater are usually the owners who operate and manage the catchment system, hence, they are more likely to exercise water conservation because they know how much water is in storage and they will try to prevent the storage tank from drying up.

v) Rainwater harvesting technologies are flexible and can be built to meet almost any requirements. Construction, operation, and maintenance are not labour intensive.

Types of Rainwater Harvesting Systems

Typically, a rainwater harvesting system consists of three basic elements: the collection system, the conveyance system, and the storage system. Collection systems can vary from simple types within a household to bigger systems where a large catchment area contributes to an impounding reservoir from which water is either gravitated or pumped to water treatment plants. The categorisation of rainwater harvesting systems depends on factors like the size and nature of the catchment areas and whether the systems are in urban or rural settings. Some of the systems are described below.

(i) Simple roof water collection systems

While the collection of rainwater by a single household may not be significant, the impact ofthousands or even millions of household rainwater storage tanks can potentially be enormous. The main components in a simple roof water collection system are the cistern itself, the piping that leads to the cistern and the appurtenances within the cistern. The materials and the degree of sophistication of the whole system largely depend on the initial capital investment. Some cost effective systems involve cisterns made with ferro-cement, etc. In some cases, the harvested rainwater may be filtered. In other cases, the rainwater may be disinfected.

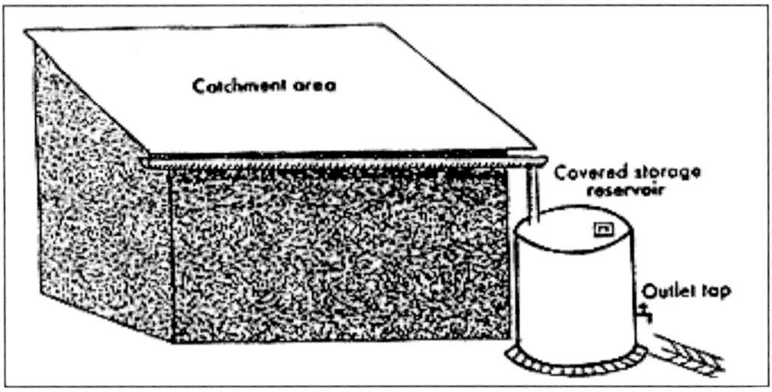

Figure 1. Example of a roof catchment system

(ii) Larger systems for educational institutions, stadiums, airports, and other facilities

When the systems are larger, the overall system can become a bit more complicated, for example rainwater collection from the roofs and grounds of institutions, storage in underground reservoirs, treatment and then use for non-potable applications.

(iii) Roof water collection systems for high-rise buildings in urbanised areas

In high-rise buildings, roofs can be designed for catchment purposes and the collected roof water can be kept in separate cisterns on the roofs for non-potable uses.

(iv) Land surface catchments

Rainwater harvesting using ground or land surface catchment areas can be a simple way of collecting rainwater. Compared to rooftop catchment techniques, ground catchment techniques provide more opportunity for collecting water from a larger surface area. By retaining the flows (including flood flows) of small creeks and streams in small storage reservoirs (on surface or underground) created by low cost (e.g., earthen) dams, this technology can meet water demands during dry periods. There is a possibility of high rates of water loss due to infiltration into the ground, and because of the often marginal quality of the water collected, this technique is mainly suitable for storing water for agricultural purposes.

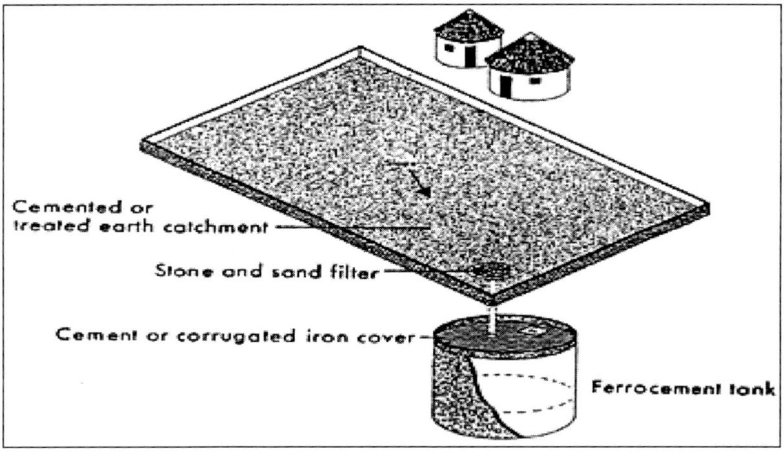

Figure 2. Example of a ground catchment system

(v) Collection of stormwater in urbanised catchments

The surface runoff collected in stormwater ponds/reservoirs from urban areas is subject to a wide variety of contaminants. Keeping these catchments clean is of primary importance, and hence the cost of water pollution control can be considerable.

Contribute to Sustainable Water Strategy

Many cities around the world obtain their water from great distances—often over 100km away. But this practice of increasing dependence on the upper streams of the water resource supply area is not sustainable. Building dams in the upper watershed often means submerging houses, fields and wooded areas. It can also cause significant socio-economic and cultural impacts in the affected communities. In addition, some existing dams have been gradually filling with silt. If not properly maintained by removing these sediments, the quantity of water collected may be significantly reduced.

When the city increases the degree of its dependence on a remote water resource, and there is a long period without rainfall in the upstream dam sites, the ability of the city to function effectively is seriously compromised. The same can be said about a city's reliance on a pipeline for drawing water from a water resource area to the city. A city which is totally reliant on a large, centralised water supply pipeline (or "life-line") is vulnerable in the face of a large-scale natural disaster. A shift from "life-line" to decentralised "life-points" should be encouraged. Numerous scattered water resource "life-points" within a city are more resilient and can draw on rainwater and groundwater, providing the city with greater flexibility in the face of water shortages and earthquakes.

Hydrological Cycle

Due to the rapid pace of urbanisation, many of the world's large cities are facing problems with urban floods. The natural hydrological cycle manifests itself at different scales, depending upon climatic, geographic and biological factors. As rain falls over time and seeps underground to become groundwater, it feeds submerged springs and rivers. The concrete and asphalt structures of cities have tended to disrupt the natural hydrological cycle, and reduce the amount of rainwater permeating underground. A decrease in the area where water can penetrate speeds up the surface flow of rainwater, causing water to accumulate in drains and streams within a short time.

Every time there is concentrated heavy rain, there is an overflow of water from drains, and small and medium sized rivers and streams repeatedly flood. These conditions can often lead to an outpouring of sewage into rivers and streams from sewer outlets and sewer pumping stations, thus contaminating the quality of urban streams and rivers.

Concrete and asphalt have a profound impact on the ecology of the city. These include:

i) *Drying of the city* – This happens as rivers and watercourses are covered, natural springs dry up, and greenery is cut down.

ii) *Heat pollution* – In some cities during the hot summer, an asphalt road at midday can reach temperatures of over 60°C. The heat expelled from air conditioners can further aggravate this.

In order to achieve a comprehensive solution to this problem, new approaches to urban development are required emphasising sustainability and the restoration of the urban hydrological cycle. Traditionally, storm sewer facilities have been developed based on the assumption that the amount of rainwater drained away will have to be increased. From the standpoint of preserving or restoring the natural water cycle, it is important to retain rainwater and to facilitate its permeation by preserving natural groundcover and greenery.

Concept of "Cycle Capacity"

"Cycle capacity" refers to the time that nature needs revive the hydrological cycle. The use of groundwater should be considered from the point of view of cycle capacity. Rain seeps underground and over time becomes shallow stratum groundwater. Then, over a very long period of time, it becomes deep stratum groundwater. For sustainable use of groundwater, it is necessary to consider the storage capacity for groundwater over time. If this is neglected and groundwater is extracted too quickly, it will disappear within a short time.

Demand Side Management of Water Supply

In establishing their water supply plans, cities have usually assumed that the future demand for water will continue to increase. Typically, city waterworks departments have made excessive estimates of the demand for water and have built waterworks infrastructure based on the assumption of continued development of water resources and strategies to enlarge the area of water supply.

The cost of development is usually recovered through water rates, and when there is plenty of water in the resource area, conservation of the resource is not promoted. This tends to create a conflict when drought occurs, due to the lack of policies and programmes to encourage water conservation. It has even been suggested that the lack of promotion of water conservation and rainwater harvesting is due to the need to recover infrastructure development costs through sales of piped water. The exaggerated projection of water demand leads to the over-development of water resources, which in turn encourages denser population and more consumption of water.

Sustainability of urban water supply requires a change from coping with water supply without controlling demand, to coping with supply by controlling demand. The introduction of demand side management encourages all citizens to adopt a water conservation approaches, including the use of freely available, locally supplied rainwater.

Quality and Health Aspects in Utilising Rainwater

In the past, it was believed that rainwater was pure and could be consumed without pre-treatment. While this may be true in some areas that are relatively unpolluted, rainwater collected in many locations contains impurities. Particularly during the last three decades, "acid rain" has affected the quality of the collected water, to the point where it now usually requires treatment. Rainwater quality varies for a number of reasons. While there are widely accepted standards for drinking water, the development of approved standards for water when it is used for non-potable applications would facilitate the use of rainwater sources.

In terms of physical-chemical parameters, collected roof water, rainwater and urban storm water tend to exhibit quality levels that are generally comparable to the World Health Organisation (WHO) guideline values for drinking water. However, low pH rainwater can occur as a result of sulphur dioxide, nitrous oxide and other industrial emissions, hence air quality standards must be reviewed and enforced. In addition, high lead values can sometimes be attributed to the composition of certain roofing materials – thus it is recommended that for roof water collection systems, the type of roofing material should be carefully considered. A number of collected rainwater samples have exceeded the WHO values in terms of total coliform and faecal coliform. The ratios of faecal coliform to faecal streptococci from these samples indicated that the source of pollution was the droppings of birds, rodents, etc.

Currently, water quality control in roof water collection systems is limited to diverting first flushes and occasional cleaning of cisterns. Boiling, despite its limitations, is the easiest and surest way to achieve disinfection, although there is often a reluctance to accept this practice as taste is affected. Chlorine in the form of household bleach can be used for disinfection, however the cost of UV disinfection systems are usually prohibitive. One promising area of research is the use of photo-oxidation based on available sunlight to remove both the coliforms and streptococci.

Design and Maintain Facilities

Catchment Surface

The effective catchment area and the material used in constructing the catchment surface influence the collection efficiency and water quality. Materials commonly used for roof catchment are corrugated aluminium and galvanised iron, concrete, fibreglass shingles, tiles, slates, etc. Mud is used primarily in rural areas. Bamboo roofs are least suitable because of possible health hazards. The materials of catchment surfaces must be non-toxic and not contain substances which impair water quality.

For example, asbestos roofs should be avoided; also, painting or coating of catchment surfaces should be avoided if possible. If the use of paint or coating is unavoidable, only non-toxic paint or coating should be used; lead, chromium, and zinc-based paints/coatings should be avoided. Similarly, roofs with metallic paint or other coatings are not recommended as they may impart tastes or colour to the collected water. Catchment surfaces and collection devices should be cleaned regularly to remove dust, leaves and bird droppings so as to minimise bacterial contamination and maintain the quality of collected water. Roofs should also be free from over-hanging trees since birds and animals in the trees may defecate on the roof.

When land surfaces are used as catchment areas, various techniques are available to increase runoff capacity, including: i) clearing or altering vegetation cover, ii) increasing the land slope with artificial ground cover, and iii) reducing soil permeability by soil compaction. Specially constructed ground surfaces (concrete, paving stones, or some kind of liner) or paved runways can also be used to collect and convey rainwater to storage tanks or reservoirs. In the case of land surface catchments, care is required to avoid damage and contamination by people and animals. If required, these surfaces should be fenced to prevent the entry of people and animals. Large cracks in the paved catchment due to soil movement, earthquakes or exposure to the elements should be repaired immediately. Maintenance typically consists of the removal of dirt, leaves and other accumulated materials. Such cleaning should take place annually before the start of the major rainfall season.

Conveyance Systems

Conveyance systems are required to transfer the rainwater collected on catchment surfaces (e.g. rooftops) to the storage tanks. This is usually accomplished by making connections to one or more down-pipes connected to collection devices (e.g. rooftop gutters). The pipes used for conveying rainwater, wherever possible, should be made of plastic, PVC or other inert substance, as the pH of rainwater can be low (acidic) and may cause corrosion and mobilisation of metals in metal pipes.

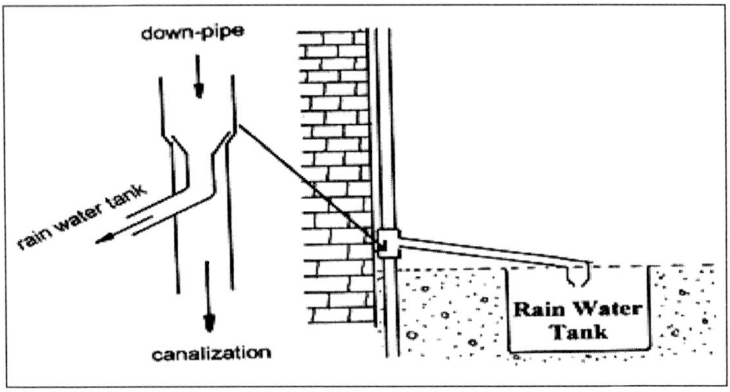

Figure 3. Example of a first flush device installation

When selecting a conveyance system, consideration should be given to the fact that when it first starts to rain, dirt and debris from catchment surfaces and collection devices will be washed into the conveyance systems (e.g. down-pipes). Relatively clean water will only be available sometime later in the storm. The first part of each rainfall should be diverted from the storage tank. There are several possible options for selectively collecting clean water for the storage tanks. The common method is a sediment trap, which uses a tipping bucket to prevent the entry of debris from the catchment surface into the tank. Installing a first flush (or foul flush) device is also useful to divert the initial batch of rainwater away from the tank.

Gutters and down-pipes need to be periodically inspected and carefully cleaned. A good time to inspect gutters and down-pipes is while it is raining, so that leaks can be easily detected. Regular cleaning is necessary to avoid contamination.

Storage Tanks

Storage tanks for collected rainwater may be located either above or below the ground. They may be constructed as a part of the building, or may be built as a separate unit located some distance away from the building. The design considerations vary according to the type of tank and other factors. Various types of rainwater storage facilities can be found in practice. Storage tanks should be constructed of inert material. Reinforced concrete, fibreglass, polyethylene, and stainless steel are suitable materials.

Ferro-cement tanks and jars made of mortar or earthen materials are commonly used. As an alternative, interconnected tanks made of pottery or polyethylene may be suitable. The polyethylene tanks are compact but have a large storage capacity (1,000 to 2,000 litres). They are easy to clean and have many openings which can be fitted with connecting pipes. Bamboo reinforced tanks are less successful because the bamboo may become infested with termites, bacteria and fungus.

Precautions are required to prevent the entry of contaminants into storage tanks. The main sources of external contamination are pollution from debris, bird and animal droppings, and insects that enter the tank. Sometimes, human, animal and other environmental contaminants, which happen to fall into tanks, can cause contamination. Open containers are not recommended for storing water for drinking purposes. A solid and secure cover is required to avoid breeding of mosquitoes, to prevent insects and rodents from entering the tank, and to keep out sunlight to prevent the growth of algae inside the tank. A coarse inlet filter is also desirable for excluding coarse debris, dirt, leaves, and other solid materials.

The storage tank should be checked and cleaned periodically. All tanks need cleaning and their designs should allow for thorough scrubbing of the inner walls and floors. A sloped bottom and the provision of a sump and a drain are useful for collection and discharge of settled grit and sediment. An entrance hole is required for easy access for cleaning. The use of a chlorine solution is recommended for cleaning, followed by thorough rinsing. Chlorination of the cisterns or storage tanks is necessary if the water is to be used for drinking and domestic uses. Dividing tanks into two sections or dual tanks can facilitate cleaning. Cracks in the storage tanks can create major problems and should be repaired immediately.

The extraction system (e.g., taps/faucets, pumps) must not contaminate the stored water. Taps/faucets should be installed at least 10 cm above the base of the tank as this allows any debris entering the tank to settle on the bottom, where if it remains undisturbed, will not affect the quality of the water. Rainwater pipes must be permanently marked in such a way that there is no risk of confusing them with drinking water pipes. Taps must also be clearly labelled for the user both in the local language and in clear graphic images. The handle of taps might be detachable to avoid the misuse by children. Periodic maintenance should also be carried out on any pumps used to lift water to selected areas in the house or building.

The following devices are also desirable.

i) An overflow pipe leading into either infiltration plants, drainage pipes with sufficient capacity or the municipal sewage pipe system.

ii) An indicator of the amount of water in the storage tank

iii) A vent for air circulation (often the overflow pipe can substitute)

iv) Protection against insects, rodents, vermin, etc. may also be required.

When using rainwater, it is important to recognise that the rainfall is not constant throughout the year; therefore, planning the storage system with an adequate capacity is required for the constant use of rainwater even during dry periods. Knowledge of the rainfall quantity and seasonality, the area of the catchment surface and volume of the storage tank, and quantity and period of use required for water supply purposes is critical. For example, in Tokyo, the average annual rainfall is about 1,400 mm. Assuming that the effective catchment area of a house is equal to the horizontal line of its roof surface area, and given that the roof surface area is 50 m^2, the average annual volume of rainwater falling on the roof may be calculated as 70 m^3. However, in practice, this volume can never be achieved since a portion of the rainwater evaporates from the roof surface and a portion may be lost to the drainage system, including the first flush. Furthermore, a portion of collected rainwater volume may be lost as overflow from the storage container if the storage tank has insufficient capacity to store the entire collected volume even in a heavy rain. Thus, the net usable or available amount of rainwater from the roof surface would be approximately 70% to 80% of the gross volume of rainfall. In the above example, the actual usable amount of rainwater would be about 49 m^3 to 56 m^3 in a year.

The disadvantages of the rainwater harvesting and utilisation systems are:

i) The catchment area and storage capacity of a system are relatively small. There is a great variation in weather. During a prolonged drought, the storage tank may dry up.

ii) Maintenance of rainwater harvesting systems, and the quality of collected water, can be difficult for users.

iii) Extensive development of rainwater harvesting systems may reduce the income of public water systems.

iv) Rainwater harvesting systems are often not part of the building code and lack clear guidelines for users/developers to follow.

v) Rainwater utilisation has not been recognised as an alternative of water supply system by the public sector. Governments typically do not include rainwater utilisation in their water management policies, and citizens do not demand rainwater utilisation in their communities.

vi) Rainwater storage tanks may be a hazard to children who play around it.

viii) Rainwater storage tanks may take up valuable space.

ix) Some development costs of larger rainwater catchment system may be too high if the costs are not shared with other systems as part of a multi-purpose network

A Systemic Approach to the Design of Rainwater Utilisation System

Rainwater utilisation, together with water conservation and wastewater reclamation, should be incorporated into municipal ordinances and regulations. Some standardisation of materials, at least at regional level, may be desirable from a maintenance and replacement point of view. It may be also appropriate to standardise the design of the rainwater utilisation system, at least at the regional level.

Implementation Policies

Various implementation policies should be established to make rainwater utilisation and other measures a part of the social system.

Leadership is very important and local governments must take the initiative to promote the concept of water resource independence and restoration of the natural hydrological cycle. Consideration should be given to subsidising facilities for rainwater utilisation.

Technology and Human Resource Development

Encouraging technology and human resources development to support rainwater utilisation is very important. It is also important to promote the development of efficient and affordable devices to conserve water, facilities to use rainwater and devices to enhance the underground seepage of rainwater. Together with this, there is a need to train specialists with a thorough grasp of these technologies and devices.

Networking

To promote rainwater harvesting and utilisation as an environmentally sound approach for sustainable urban water management, a network should be established involving government administrators, citizens, architects, plumbers and representatives of equipment manufacturers. It is essential to encourage regional exchanges amongst public servants, citizens and industry representatives involved in rainwater storage, seepage and use, as well as the conservation and reclamation of water.

Public awareness and education are essential to improve acceptance and implementation of rainwater harvesting and utilisation.

Examples of Rainwater Harvesting and Utilisation

In Singapore, which has limited land resources and a rising demand for water, is on the lookout for alternative sources and innovative methods of harvesting water. Almost 86% of Singapore's population lives in high-rise buildings. A light roofing is placed on the roofs to act as catchment. Collected roof water is kept in separate cisterns on the roofs for non-potable uses. A recent study of an urban residential area of about 742 ha used a model to determine the optimal storage volume of the rooftop cisterns, taking into consideration non-potable water demand and actual rainfall at 15-minute intervals. This study demonstrated an effective saving of 4% of the water used, the volume of which did not have to be pumped from the ground floor. As a result of savings in terms of water, energy costs, and deferred capital, the cost of collected roof water was calculated to be S$0.96 against the previous cost of S$1.17 per cubic meter.

A marginally larger rainwater harvesting and utilisation system exists in the Changi Airport. Rainfall from the runways and the surrounding green areas is diverted to two impounding reservoirs. One of the reservoirs is designed to balance the flows during the coincident high runoffs and incoming tides, and the other reservoir is used to collect the runoff. The water is used primarily for non-potable functions such fire-fighting drills and toilet flushing. Such collected and treated water accounts for 28 to 33% of the total water used, resulting in savings of approximately S$ 390,000 per annum.

In Tokyo, rainwater harvesting and utilisation is promoted to mitigate water shortages, control floods, and secure water for emergencies. The Ryogoku Kokugikan Sumo-wrestling Arena, built in 1985 in Sumida City, is a well-known facility that utilises rainwater on a large scale. The 8,400 m^2 rooftop of this arena is the catchment surface of the rainwater utilisation system. Collected rainwater is drained into a 1,000 m^3 underground storage tank and used for toilet flushing and air conditioning. Sumida City Hall uses a similar system. Following the example of Kokugikan, many new public facilities have begun to introduce rainwater utilisation systems in Tokyo.

At the community level, a simple and unique rainwater utilisation facility, "Rojison", has been set up by local residents in the Mukojima district of Tokyo to utilise rainwater collected from the roofs of private houses for garden watering, fire-fighting and drinking water in emergencies. To date, about 750 private and public buildings in Tokyo have introduced rainwater collection and utilisation systems. Rainwater utilisation is now flourishing at both the public and private levels.

In October 1998, rainwater utilisation systems were introduced in Berlin as part of a large scale urban re-development, the DaimlerChrysler Potsdamer Platz, to control urban flooding, save city water and create a better micro climate. Rainwater falling on the rooftops (32,000 m^2) of 19 buildings is collected and stored in a 3500 m^3 rainwater basement tank. It is then used for toilet flushing, watering of green areas (including roofs with vegetative cover) and the replenishment of an artificial pond.

In another project at Belss-Luedecke-Strasse building estate in Berlin, rainwater from all roof areas (with an approximate area of 7,000 m^2) is discharged into a separate public rainwater sewer and transferred into a cistern with a capacity of 160 m^3, together with the runoff from streets, parking spaces and pathways (representing an area of 4,200 m^2). The water is treated in several stages and used for toilet flushing as well as for garden watering. The system design ensures that the majority of the pollutants in the initial flow are flushed out of the rainwater sewer into the sanitary sewer for proper treatment in a sewage plant. It is estimated that 58% of the rainwater can be retained locally through the use of this system. Based on a 10-year simulation, the savings of potable water through the utilisation of rainwater are estimated to be about 2,430 m^3 per year, thus preserving the groundwater reservoirs of Berlin by a similar estimated amount.

Both of these systems not only conserve city water, but also reduce the potential for pollutant discharges from sewerage systems into surface waters that might result from stormwater overflows. This approach to the control of non point sources of pollution is an important part of a broader strategy for the protection of surface water quality in urban areas.

Storing rainwater from rooftop run-off in jars is an appropriate and inexpensive means of obtaining high quality drinking water in Thailand. Prior to the introduction of jars for rainwater storage, many communities had no means of protecting drinking water from waste and mosquito infestation. The jars come in various capacities, from 100 to 3,000 litres and are equipped with lid, faucet, and drain. The most popular size is 2,000 litres, which costs 750 Baht, and holds sufficient rainwater for a six-person household during the dry season, lasting up to six months.

Two approaches are used for the acquisition of water jars. The first approach involves technical assistance and training villagers on water jar fabrication. This approach is suitable for many villages, and encourages the villagers to work cooperatively. Added benefits are that this environmentally appropriate technology is easy to learn, and villagers can fabricate water jars for sale at local markets. The second approach is applicable to those villages that do not have sufficient labour for making water jars. It involves access to a revolving loan fund to assist these villages in purchasing the jars. For both approaches, ownership and self-maintenance of the water jars are important. Villagers are also trained on how to ensure a safe supply of water and how to extend the life of the jars.

Initially implemented by the Population and Community Development Association (PDA) in Thailand, the demonstrated success of the rainwater jar project has encouraged the Thai government to embark on an extensive national program for rainwater harvesting.

In Indonesia, groundwater is becoming more scarce in large urban areas due to reduced water infiltration. The decrease of groundwater recharge in the cities is directly proportional to the increase in the pavement and roof area. In addition, high population density is has brought about high groundwater consumption. Recognising the need to alter the drainage system, the Indonesian government introduced a regulation requiring that all buildings have an infiltration well. The regulation applies to two-thirds of the territory, including the Special Province of Yogyakarta, the Capital Special Province of Jakarta, West Java and Central Java Province. It is estimated that if each house in Java and Madura had its own infiltration well, the water deficit of 53% by the year of 2000 would be reduced to 37%, which translates into a net savings of 16% through conservation.

In the Philippines, a rainwater harvesting programme was initiated in 1989 in Capiz Province with the assistance of the Canadian International Development Research Centre (IDRC). About 500 rainwater storage tanks were constructed made of wire-framed ferro-cement, with capacities varying from 2 to 10 m^3. The construction of the tanks involved building a frame of steel reinforcing bars (rebar) and wire mesh on a sturdy reinforced concrete foundation. The tanks were then plastered both inside and outside, thereby reducing their susceptibility to corrosion relative to metal storage tanks.

The rainwater harvesting programme in Capiz Province was implemented as part of an income generation initiative. Under this arrangement, loans were provided to fund the capital cost of the tanks and related agricultural operations. Loans of US$200, repayable over a three-year period, covered not only the cost of the tank but also one or more income generating activities such as the purchase and rearing of pigs, costing around US$25 each. Mature pigs can sell for up to US$90 each, providing an income opportunity for generating that could provide sufficient income to repay the loan. This type of innovative mechanism for financing rural water supplies can help avoid the requirement for water resources development subsidies.

In Bangladesh, rainwater collection is seen as a viable alternative for providing safe drinking water in arsenic affected areas. Since 1997, about 1000 rainwater harvesting systems have been installed in the country, primarily in rural areas, by the NGO Forum for Drinking Water Supply & Sanitation. This Forum is the national networking and service delivery agency for NGOs, community-based organisations and the private sector concerned with the implementation of water and sanitation programmes in unserved and underserved rural and urban communities. Its primary objective is to improve access to safe, sustainable, affordable water and sanitation services and facilities in Bangladesh.

The rainwater harvesting tanks in Bangladesh vary in capacity from 500 litres to 3,200 litres, costing from Tk. 3000-Tk.8000 (US$ 50 to US$ 150). The composition and structure of the tanks also vary, and include ferro-cement tanks, brick tanks, RCC ring tanks, and sub-surface tanks. The rainwater that is harvested is used for drinking and cooking and its acceptance as a safe, easy-to-use source of water is increasing amongst local users. Water quality testing has shown that water can be preserved for four to five months without bacterial contamination. The NGO Forum has also undertaken some recent initiatives in urban areas to promote rainwater harvesting as an alternative source of water for all household purposes.

Gansu is one of the driest provinces in China. The annual precipitation is about 300 mm, while potential evaporation amounts to 1500-2000 mm. Surface water and groundwater is limited, thus agriculture in the province relies on rainfall and people generally suffer from inadequate supplies of drinking water. Since the 1980s, research, demonstration and extension projects on rainwater harvesting have been carried out with very positive results.

In 1995/96, the "121" Rainwater Catchment Project implemented by the Gansu Provincial Government supported farmers by building one rainwater collection field, two water storage tanks and providing one piece of land to grow cash crops. This project has proven successful in supplying drinking water for 1.3 million people and developing irrigated land for a courtyard economy. As of 2000, a total of 2,183,000 rainwater tanks had been built with a total capacity of 73.1 million m^3 in Gansu Province, supplying drinking water for 1.97 million people and supplementary irrigation for 236,400 ha of land.

Rainwater harvesting has become an important option for Gansu Province to supply drinking water, develop rain-fed agriculture and improve the ecosystem in dry areas. Seventeen provinces in China have since adopted the rainwater utilisation technique, building 5.6 million tanks with a total capacity of 1.8 billion m^3, supplying drinking water for approximately 15 million people and supplemental irrigation for 1.2 million ha of land.

Although in some parts of Africa rapid expansion of rainwater catchment systems has occurred in recent years, progress has been slower than Southeast Asia. This is due in part to the lower rainfall and its seasonal nature, the smaller number and size of impervious roofs and the higher costs of constructing catchment systems in relation to typical household incomes. The lack of availability of cement and clean graded river sand in some parts of Africa and a lack of sufficient water for construction in others, add to overall cost.

Nevertheless, rainwater collection is becoming more widespread in Africa with projects currently in Botswana, Togo, Mali, Malawi, South Africa, Namibia, Zimbabwe, Mozambique, Sierra Leone and Tanzania among others. Kenya is leading the way. Since the late 1970s, many projects have emerged in different parts of Kenya, each with their own designs and implementation strategies. These projects, in combination with the efforts of local builders called "fundis" operating privately and using their own indigenous designs, have been responsible for the construction of many tens of thousands of rainwater tanks throughout the country. Where cheap, abundant, locally available building materials and appropriate construction skills and experience are absent; ferro-cement tanks have been used for both surface and sub-surface catchment.

Due to inadequate piped water supplies, the University of Dar es Salaam, Tanzania has applied rainwater harvesting and utilisation technology to supplement the piped water supply in some of the newly built staff housing. Rainwater is collected from the hipped roof made with corrugated iron sheets and led into two "foul" tanks, each with a 70-litre capacity. After the first rain is flushed out, the foul tanks are filled up with rainwater. As the foul tanks fill up, settled water in the foul tanks flows to two underground storage tanks with a total capacity of 80,000 litres.

Then, the water is pumped to a distribution tank with 400 litres capacity that is connected to the plumbing system of the house. The principles for the operation of this system are: (i) only one underground tank should be filled at a time; (ii) while one tank is being filled, water can be consumed from the other tank, (iii) rainwater should not be mixed with tap water; (iv) underground storage tanks must be cleaned thoroughly when they are empty; (v) in order to conserve water, water should only be used from one distribution tank per day.

Thousands of roof catchment and tank systems have been constructed at a number of primary schools, health clinics and government houses throughout Botswana by the town and district councils under the Ministry of Local Government, Land and Housing (MLGLH). The original tanks were prefabricated galvanised steel tanks and brick tanks. The galvanised steel tanks have not performed well, with a short life of approximately 5 years. The brick tanks are unpopular, due to leakage caused by cracks, and high installation costs. In the early 1980s, the MLGLH replaced these tanks in some areas with 10-20 m^3 ferro-cement tanks promoted by the Botswana Technology Centre. The experience with ferro-cement tanks in Botswana is mixed; some have performed very well, but some have leaked, possibly due to poor quality control.

Over the past decade, many NGOs and grassroots organisations have focused their work on the supply of drinking water using rainwater harvesting, and the irrigation of small-scale agriculture using sub-surface impoundments. In the semi-arid tropics of the north-eastern part of Brazil, annual rainfall varies widely from 200 to 1,000 mm, with an uneven regional and seasonal rainfall pattern. People have traditionally utilised rainwater collected in hand-dug rock catchments and river bedrock catchments.

To address the problem of unreliable rural drinking water supply in north-eastern Brazil, a group of NGOs combined their efforts with government to initiate a project involving the construction of one million rainwater tanks over a five year period, with benefits to 5 million people. Most of these tanks are made of pre-cast concrete plates or wire mesh concrete. Rainwater harvesting and utilisation is now an integrated part of educational programs for sustainable living in the semi-arid regions of Brazil. The rainwater utilisation concept is also spreading to other parts of Brazil, especially urban areas. A further example of the growing interest in rainwater harvesting and utilisation is the establishment of the Brazilian Rainwater Catchment Systems Association, which was founded in 1999 and held its 3rd Brazilian Rainwater Utilisation Symposium in the fall of 2001.

The island of Bermuda is located 917 km east of the North American coast. The island is 30 km long, with a width ranging from 1.5 to 3 km. The total area is 53.1 km^2. The elevation of most of the land mass is less than 30 m above sea level, rising to a maximum of less than 100 m. The average annual rainfall is 1,470 mm. A unique feature of Bermuda roofs is the wedge-shaped limestone "glides" which have been laid to form sloping gutters, diverting rainwater into vertical leaders and then into storage tanks.

Most systems use rainwater storage tanks under buildings with electric pumps to supply piped indoor water. Storage tanks have reinforced concrete floors and roofs, and the walls are constructed of mortar-filled concrete blocks with an interior mortar application approximately 1.5 cm thick. Rainwater utilisation systems in Bermuda are regulated by a Public Health Act which requires that catchments be whitewashed by white latex paint; the paint must be free from metals that might leach into water supplies. Owners must also keep catchments, tanks, gutters, pipes, vents, and screens in good repair. Roofs are commonly repainted every two to three years and storage tanks must be cleaned at least once every six years.

St. Thomas, US Virgin Islands, is an island city which is 4.8 km wide and 19 km long. It is situated adjacent to a ridge of mountains which rise to 457 m above sea level. Annual rainfall is in the range of 1,020 to 1,520 mm. A rainwater utilisation system is a mandatory requirement for a residential building permit in St. Thomas. A single-family house must have a catchment area of 112 m^2 and a storage tank with 45 m^3 capacity. There are no restrictions on the types of rooftop and water collection system construction materials.

Many of the homes on St. Thomas are constructed so that at least part of the roof collects rainwater and transports it to storage tanks located within or below the house. Water quality test of samples collected from the rainwater utilisation systems in St. Thomas found that contamination from faecal coliform and Hg concentration was higher than EPA water quality standards, which limits the use of this water to non-potable applications unless adequate treatment is provided.

At the U.S. National Volcano Park, on the Island of Hawaii, rainwater utilisation systems have been built to supply water for 1,000 workers and residents of the park and 10,000 visitors per day. The Park's rainwater utilisation system includes the rooftop of a building with an area of 0.4 hectares, a ground catchment area of more than two hectares, storage tanks with two reinforced concrete water tanks with 3,800 m^3 capacity each, and 18 redwood water tanks with 95 m^3 capacity each. Several smaller buildings have their own rainwater utilisation systems as well. A water treatment and pumping plant was built to provide users with good quality water.

References

Agarwal Anil, Narain Sunita eds., "Dying Wisdom, Rise, Fall and Potential of India's Traditional Harvesting Systems, State of India's Environment", *A Citizens' Report* 4, Delhi, Centre for Science and Environment, 404 p., 1997.

Ballabh Vishwa, Choudhary Kameshwar, *Consultation Workshop on Small Water Harvesting Structures: Technology, Institution and Management*, Anand, Institute of Rural Management, 44 p., 1999.

Cullis Adrian, Pacey Arnold, *Rain Water Harvesting,* London, Intermediate Technology Publisher, 1991.

Ladda, Brijgopal, *Planning and Financing Urban Water Supply: A Case Study of Ahmedabad,* Ahmedabad, CEPT, 1995.

Shankari Uma, Shah Esha, *Water Management Traditions in India*, Madras, PPST Foundations, 193 p., 1993.

CHAPTER 6

Water Harvesting as the Mode of Climate Change

As culture and climate are interlinked, there is a general human response to adapt and mitigate the sufferings associated with such climate extremes. Societal and cultural responses to prolonged drought include population dislocation, dwelling abandonment, widespread migration and state collapse. However, climate and culture can interact in many ways. Rather than migration, people may also resort to modify the dwelling environments by adapting strategies to optimise the utility of available water by harvesting rain. Where there is water on earth, virtually no matter what the physical conditions, there is life.

Water has been important for the development of cultural complexity in human society during the Holocene and earlier. Human ancestors have always used aquatic resources to their benefit, as we see the earliest association of hominid ancestors with lakes and pools dating back to 6 and 7 m.y. ago from northern Chad, central Africa. The earliest known hominids, Sahelanthropus tchadensis, may have exploited a significant aquatic fauna such as fish, crocodiles and amphibious mammals, along with primates, rodents, elephants, equids and bovids associated with gallery forest and savannah.

Natural springs, river valleys and coasts have remained favoured locations for both hominidsand modern humans. Particularly, aquatic and maritime adaptations such as coastal occupation, fishing and seafaring played a significant role in demographic as well as geographic expansion of modern humans after about 150,000 years ago.

A comprehensive knowledge of climate fluctuations and corresponding adaptation by human society is crucial for our progress towards sustainability. Since the adaptation through rainwater collection may be particularly effective in tropical monsoon regions, where the seasonal cycle in rainfall is large.

Holocene Climate

For a long time, climate of the Holocene epoch (~ 11 kyr BP to present), compared to last glacial period, was considered to be stable. But that belief has been broken as detailed and well-dated worldwide palaeoclimate records have provided evidence of climate variability and extreme climate events such as aridity, drought, flood, cyclone, stormy rainfall and millennial-scale variation in climate during the Holocene. The most recent of these events was the Little Ice Age (LIA), which occurred during 1300-1850, reducing the subtropical sea-surface temperatures by 3° to 4°C.

Indeed, pace of the Holocene events and of abrupt climate shifts during the last glaciation has been suggested to be statistically the same; together, they make up a series of climate shifts with a cyclicity close to 1470 ± 500 years. These changes may have influenced some of the most momentous revolutions—for instance, evolution of cultural diversity, domestication of medicinal plants, crops and animals, and adaptive innovations through socio-economic development—in human history during the Holocene.

Shifts in early Holocene atmospheric CO_2 concentration and abrupt early to mid-Holocene climatic transition have been registered both at the equator and the poles. Holocene climate fluctuations as inferred from proxy records worldwide, including North Atlantic, western Canada, southwest Ireland, North America and elsewhere are now well established. Records of sedimentation in Laguna Pallcacocha, southern Ecuador suggest El Nino/Southern Oscillation (ENSO) variability on millennial timescales throughout the Holocene. It is notable that El Nino events influenced prehistoric cultural development in Peru.

The onset of El Nino events at 5.8 kyr ago is coincident with the beginning of a monumental temple construction on the Peruvian coast, and the increase in El Nino frequency after 3.2-2.8 kyr ago correlates with the abandonment of monumental temples in the region, In addition, several extreme arid events and droughts have been documented in Asia , Africa , North America , and South America, These may represent individual episodes in a recurrent pattern of dry events, but some events may be consistent with greenhouse gas forcing. For instance, failure of the winter and spring rains and increased temperatures led to drought during 1998-2002 over the Northern Hemisphere mid-latitudes across the United States, the Mediterranean, southern Europe, and Southwest and Central Asia. These droughts were linked through a common oceanic influence and were consistent with greenhouse gas forcing,

Climate fluctuations that are not necessarily anthropogenic. Some proxies representing local climate at different places across the world reveal that the 20th century is probably neither the warmest nor a uniquely extreme climatic period of the last millennium, A variable sun through changes in solar irradiance and processes such as biological regime shifts may have caused the observed climate variability and climate fluctuations over much of the Holocene. Indeed, geophysical, archaeological and historical evidences support a solar output model for climate change during a large part of the Holocene. As inferred from biological and geological proxies, terrestrial palaeotemperature may have been higher due to natural variability.

Just before the Holocene, SSTs increased by 3.5-4.0°C during the last two glacial-interglacial transitions, synchronous with the global increase in atmospheric CO_2 and Antarctic warming, but the temperature increase occurred 2000-3000 years before the Northern Hemisphere ice sheets melted, Deglaciation lagged the melting of ice sheets in the Northern Hemisphere by millennia and may still be under way, and may remain so in the future. Thus, before the twentieth century anthropogenic climate change and its impacts that we are witnessing, the last truly global climate change occurred between 20,000 and 10,000 years ago.

At the last Glacial Interglacial Transition rise in tropical temperature was felt first, for example in the Andes and the Indo-Pacific warm pool, propagation of warming signal from the tropics to drive ice-sheet melting in the Northern Hemisphere 2000 to 3000 years later that resulted in global rise in sea level of 120 m. There was a rise of atmospheric CO_2 by nearly 90 parts per million (p.p.m.), from a low of about 180 p.p.m. in glacial times and parts of the earth warmed up by 10-12°C.

Climate Change and Human Migration

The archaeological and historical records show many instances of societal collapse, attributed to a combination of social, political and economic factors. Climate change and drought as the primary agent in human migration, cultural separation, population dislocation and the collapse of prehistoric and early historic societies. Human migration due to climate change and environmental stress is a well resolved survival strategy across continents, including Africa , Eurasia, South America and Australia. Ice cores from Kilimanjaro provide an ~ 11.7 kyr record of Holocene climate variability for eastern equatorial Africa, The periods of abrupt climate change shown in ice cores include ~ 8.3, ~ 5.2 kyr ago, and a third period, ~ 4 kyr ago coinciding with the 'First Dark Age' known for the most severe drought in tropical Africa.

In a more recent history, effects of variable precipitation and drought in equatorial east Africa during the past 1100 years bear links with climate change and cultural effects. Oral traditions as well as scientific evidence both indicate drought-induced famine, political unrest and large-scale migration of native population, and establish strong connection between cultural development, climate change and water stress. Initial human occupation coincided with a change from very dry phase to a humid phase. Archaic open campsites at elevations above 3600 m show that humans lived around late Glacial/early Holocene palaeolakes on the Altiplano.

The fact that climate change was instrumental for human occupation and abandonment of the area also provides support to proposition that early people in the Atacama region focused on favourable, humid habitats such as lakes, springs and streams, History tells us that cultures do not give up until they have exhausted options for survival over the area they occupied for longer period.

A 10,000-year simulation study over Mexico suggests that a 'megadrought' and thirteen other devastating droughts lasting over a decade resulted in 20 to 40% reduction in rainfall and occurred with a return period of less than 1000 years. The Mayan civilisation was developed around 3000 years ago in Mesoamerica , and faced recurrent droughts due to solar forcing before it collapsed due to climate deterioration towards the end of the Classic Period. A recent study has shown that the collapse of the Mayan civilisation in the Terminal Classic Period occurred during a spatio-temporally extensive dry period in the region, interspersed by more severe droughts centred at about AD 810, 860, and 910.

Ancient reservoir technology developed by the Mayan people in the seasonally dry tropics of southern Maya lowlands reveals that rainwater storage was a major source of water supply during the dry season. Reservoirs were constructed, for example, in Tikal to cope with seasonal scarcity of water. There were at least three distinct types of reservoirs as: centrally located reservoirs, residential reservoirs and margin reservoirs. The nomenclature is essentially based on location of the reservoir and the differential capacity to store water. Similar water collection and storage have been documented in other areas of South America as well.

Ancient man-made earthworks of Bolivian Amazon are another example of adaptation. The crucial factor that facilitated early domestication of palms and harvest of fish was construction of large-scale earthworks such as documented by Erickson in the Bolivian Amazon. A complex man-made network of hydraulic systems and fish weirs covering 525 km^2 in the ancient landscape of Baures region of Bolivia is a classic example of rainwater collection that provided sufficient animal protein and palm fruits to feed large and dense populations in an arid savanna.

There are numerous studies that have documented abrupt climate change, extreme drought conditions and storminess over North America during the Holocene. The 'Dust Bowl' drought severely impacted USA during 1934, 1936, and 1939-40. Recent studies of longer term US Great Plains drought variability over the past 2000 years with the use of palaeoclimatic data imply that no droughts as severe as those of the 1930s have occurred since the 1700s. A particularly notable example, supported by tree-ring climatic data, is the Lost Colony of Roanoke Island, which disappeared during the most extreme drought in 800 years.

Increasing depth of wells through time found at archaeological sites in Hili near Al Ain, United Arab Emirates, is in concordance with increasing aridity and consequent decline in groundwater levels for the past 4500 years. In southern Arabia's highland Southern Jol, McCorriston and Oches have documented and dated early water impounding structures, with important implications for prehistoric Arabian resource management and food production. In the Wadi Sana and its Wadi Shumlya tributary, checkdams (water-impounding structures) have been discovered buried in silts deposited 13,000-5000 years ago. While use of checkdams for agricultural activities in this region cannot be ruled out, the Wadi Sana archaeological remains of water storage adaptation are older than known ancient Arabian agriculture.

Holocene climate fluctuations, including large spatial variation in Holocene monsoon and temperature over India are now well-resolved from various climate change proxies. Folk sayings such as 'capture rain where it rains' may have originated in response to increased aridity in the Indian region over the last few millennia. Such climate fluctuations may have given rise to traditional village tanks, ponds and earthen embankments numbering more than 1.5 million, that still harvest rainwater in 660,000 villages in India and encourage growth of vegetation in commons and agroecosystems.

References

Bond, G. *et al.*, "A pervasive millennial-scale cycle in North Atlantic Holocene and glacial climates" *Science*, 278, 1257–1266, 1997.

Chakraborti, D. *et al.*, "Arsenic groundwater contamination and sufferings of people in Rajnandgaon district, Madhya Pradesh, India", *Curr. Sci.*, 77, 502–504, 1999.

Gupta, A. K., Anderson, D. M. and Overpeck, J. T., "Abrupt changes in the Asian southwest monsoon during the Holocene and their links to the North Atlantic Ocean", *Nature*, 421, 354–357, 2003.

Hoerling, M. and Kumar, A., "The perfect ocean for drought", *Science*, 299, 691–694, 2003.

Pandey, Deep Narayan, Anil K. Gupta and David M. Anderson, "Rainwater harvesting as an adaptation to climate change", *Review article,* Current science, Vol. 85, No. 1, 10 July, 2003.

Sandweiss, D. H. *et. al.*, "Variation in Holocene El Nino frequencies: Climate records and cultural consequences in ancient Peru", *Geology*, 29, 603–606, 2001.

Scarborough, V. L. and Gallopin, G. G., "A water storage adaptation in Maya lowlands", *Science*, 251, 658–662, 1991.

CHAPTER 7

Local Water Harvesting

Water is found generally either at the extreme top (in the form of snow and glaciers) or in deep valley bottoms, while most settlements are in between without regular water supply systems for drinking, food production, and other uses. These settlements are totally dependent on rainfall and rechannelled, scarce local water resources. Water, although a fundamental need for human survival, has already become scarce for the people in these mountains, and women and children are spending most of their time fetching water for household uses. The problem of water availability is more acute in the low rainfall/rainshadow areas.

Recent studies showed that pollution of local water is also increasing. With the population doubling almost every 35 years, the problem of water supplies for mountain households will be critically acute in the coming years. This crisis of decreasing availability of water is not only going to hamper seriously the economic development of the region but is also likely to threaten the very survival of the already marginalised rural mountain people who face acute and mass poverty and have hardly any capacity to cope with such a crisis.

In China, appropriate programmes and policy incentives for harvesting rainwater and local water sources at the household and small community levels can not only boost food production, but can also help local people to increase their incomes within a very short time. Year-round availability of water through water-harvesting programmes for domestic use will also greatly reduce the drudgery of women and children who, otherwise, have to walk long distances up and down the mountain tracks to fetch water.

Climate, culture, and cultivation are critical determinants in meeting the needs for water. The challenge lies in reviving the traditional wisdom of the local people and their water management systems and using them as a basis to develop and impart efficient local water-harvesting and management systems.

The Hindu Kush-Himalayas (HKH) are home to nearly 150 million people and are the largest storehouse of fresh water outside the poles. These tallest mountains are the sources of six of the major rivers of the world and provide sustenance to nearly 500 million people inhabiting these river basins. However, much will depend on national commitment to and initiative in action and research on water harvesting and improved understanding of climate change impacts on the water resources of the region to face the challenge of the looming water crisis in the HKH mountains.

Water from fog offers an alternative source of portable water. The goal of the Nepal Water from Fog Project is to improve access to portable water for remote villages in Nepal. The Project has two main components: to assess the potential for fog water collection technology in the Nepalese environmental context and, to implement full scale, community-based fog collection projects where they are feasible.

In most Hindu Kush-Himalayan (HKH) countries, water scarcity has become a limitation for social and economic development and the situation will worsen if nothing is done immediately to improve the water supply. In the past, the emphasis was on the development of surface and groundwater resources. These sources are becoming exhausted in many areas. Shortage of surface water and fast depletion of the groundwater table has reached an alarming situation in many countries. Afghanistan, Pakistan, China, and India are already feeling the severity of the problem. China has taken a lead in using the third kind of water resource, i.e., rainwater, which is renewable and has tremendous potential to meet the growing needs in future.

Advantages of Local Water Harvesting

Rainwater catchment utilisation (RWCU) has certain advantages over surface or groundwater development. Some of these are stated below:

i) The cost of RWCU water is only 1/6 to 1/8 of water brought a long distance haul by truck.

ii) The investment for a RWCU-irrigation system can be repaid within 4-5 years.

iii) Decrease in runoff due to water exhaust by RWCU will only be 0.1-0.5% of total river flow.

iv) Rainwater is usually less polluted than surface or groundwater.

v) Rainfall is usually widely distributed and is generally available close to home unlike river water or deep groundwater.

vi) Rainwater projects are simple. They can be built by the users themselves with some input from outside.

Types of Local Water Harvesting Systems

The different types of rainwater catchment/collection systems include the following.

i) Rooftop

ii) Concrete lined surface

iii) Cement soil lined surface

iv) Plastic film, both exposed and covered with soil

v) Compacted soil

vi) Roads/runways/paved surfaces

Types of Storage Sub-Systems

i) Water cellar

ii) Water tank

iii) Water jar/cistern (above and underground)

iv) Pond/reservoir

Increasing Soil Water Use Efficiency

Measures to increase the efficiency of soil water use in rainfed agriculture include the following.

i) Deep ploughing

ii) Harrowing and raking

iii) Soil surface plastic covering to reduce evaporation losses

iv) Inner ridge ditch culture and ridge culture covered with plastic to concentrate runoff

v) Terracing

vi) Contour line ditching,

vii) Fish scale pits (micro-catchments)

Optimum and efficient use of harvested water is an integral part of ensuring that water-harvesting systems are adaptable and sustainable. This includes the selection of right types of crops, e.g., orchards and vegetables, and irrigation systems with greater efficiency. It is also advisable to limit the use of stored water to drinking, cooking, and dish washing when the water level in the domestic storage system is not replenished. Some of the efficient irrigation systems and associated technologies include:

i) Drip irrigation
ii) Micro-sprinklers
iii) Seepage irrigation
iv) Point irrigation

As demand for water grows exponentially all available supplies will not be adequate, giving rise to rationing, pricing, and even loss of supply in extreme cases. An interesting example of how water needs and supply conditions are changing can be seen from a tourist resort area outside Kathmandu.

Government and, lately, non-government organisations have played a major role in building infrastructure and providing public services. External interventions in local water harvesting have focussed on developing piped drinking water and irrigation systems.

External intervention in local development areas has both desirable and undesirable impacts. Over a couple of decades, significant numbers of people in rural areas have access to piped water for drinking. Similarly, external assistance in small-scale hill irrigation schemes has made significant impact on agricultural productivity and cropping intensity in irrigated areas. These changes have had direct benefits on the local economy. Similarly, external interventions in irrigation schemes have helped farmers to grow two rice crops and wheat. Irrigation has reinforced the tendencies of the local farmers to keep buffaloes for milk production.

External intervention in local water harvesting (piped water) has significantly reduced the time needed to fetch water. The most undesirable impact of intervention is increased dependency on external support in local water harvesting. In settlements located in Naldung potential sources of water to build piped water and irrigation schemes exist. As external agencies have not come up with funds, such potential schemes have not been implemented.

Although it is premature to evaluate the impacts of the micro-hydel project, the likelihood of undesirable impacts is minimal. External interventions and community initiatives and effort are neatly blended to dispel undesirable impacts. External interventions through Government and NGO involvement in water harvesting in the watershed area have played different roles. In the past, the local people's interest and capacity to assume responsibility were seldom incorporated into the planning process. The focus has been on success in terms of physical completion of the project, ignoring the operation and maintenance aspects.

Water resource systems, like other CPRs, have multiple use and competing users. For instance, the Ladakhis revere their streams and maintain their cleanliness. Water, for Ladakhis, is not only important from the economic perspective alone, but is also combined with social and religious sanctions and incentives for management. In Ratura farmer-managed irrigation systems, the farmers gather before the Goddess Chandi built near the head of the canal to decide on maintenance chores and rotation sequences. Any violation of the rules fixed in the Temple is unimaginable. The claim over resources through rights is not static or stable, but is negotiated by a combination of social, political, and ecological claim- making strategies.

Another challenge in reviving the traditional institutions is the informal existence of these institutions. They do not function with a set of rules in use. Studies elsewhere indicate that, behind every formal rule of water rights' allocation in place, there is a complex network of 'working rules', through stealing and selling. This in some way makes water allocation more equal and flexible by providing means for those with limited rights to obtain water. This leads to reconstitution, renegotiation, and redefinition of socially-acceptable forms of more equitable access.

Community-based resource management systems were efficient in that they followed rules and patterns of behaviour geared towards self-sustainability. This means that traditional communities evolved mechanisms to address the issues of environment and development. In fact, these societies do not conceive of their natural environment nor of their own relations with it, as modern societies do. They perceive that surrounding natural resources are provided as an act of kindness by some supernatural agents and shortage of resources is perceived to be a temporary phenomenon.

High altitude mountainous areas that cannot be used as a land resource are available as sources of water for the mountains and plains. The different sources of water in Pakistan are used to place

17.20 mha or roughly 77% of the cultivated area, under either canal, tubewell, or tank irrigation. The remaining 23% of the area is under rainfed/snowfed systems. This does not include those mountainous areas in which no cultivation is contemplated and which, together with the rainfed/snowfed areas, constitute 17.76 mha. Agriculture and household water harvesting technologies go hand in hand in mountainous areas. Hence, in the mountains, sustenance policies must have a two-pronged strategy: water harvesting for the household and for agricultural production. The existing water-harvesting technologies for household and agricultural use are inter-related.

Development of Micro-water Conservancy Works

Micro-engineering for water conservancy works is mostly used to resolve irrigation problems in hilly regions susceptible to droughts where other types of water conservancy works are not available. Micro-water conservancy works have the advantages of low costs and quick rewards and are warmly accepted by farmers. The several years' practice of micro-water conservancy works show that the works are valuable measures for irrigating crops on otherwise dry hill slopes.

Owing to water conservancy works, a great deal of economic development has taken place in irrigated regions while there has been little change in non-irrigated regions. After three years' demonstration, government leaders, engineers, and farmers all realised that water shortage is the main factor restricting economic development in the hilly areas of Deyang and agreed that micro-water conservancy works are the only way to resolve water shortage problems in the area.

Households are usually responsible for construction and management of micro-water conservancy works. Yet governments too must strengthen their functions of organising and guiding households to construct micro-water harvesting works in a systematic manner. Construction headquarters for micro-water conservancy works were set up in the cities, counties, and villages, where the works were planned under the charge of a deputy governor.

Micro-water conservancy works are simple engineering works for engineers but not for most of the farmers. During the trial period for micro- water conservancy works, it was considered that the works had a large number of small-scale cisterns and that individual households should be responsible for construction and management of the works. The policies on micro-water conservancy works are that the benefits must go to the households responsible for construction and management. If the works are private, the local government issues the owner an ownership with certificate after the works are completed and checked by the headquarters. The owner has the right to use, transfer, and sell his/her micro-water harvesting works.

References

Agnihotri, Y.; Samra, J.S.;Mittal, S.P., "Boosting Hill Resource Economy through Watershed Management in Hoshiarpur, Shiwaliks, India", *Indian-Journal of Soil Conservation*, 24: 3, pp 248-252, 1996.

Aslam, M.; Ikram, M. Z., 'Water Harvesting and Dugwells for Sustainable Water Use', In *Progressive Farming*, 14 (5) 66-69, 1994.

United Nations Environment Programme / Tycooly International Publishing Ltd., Rain and Stormwater Harvesting in Rural Areas, 238pp. *Water Resources Series*, 5. Nairobi and Dublin: United Nations Environment Programme / Tycooly International Publishing Ltd., 1983.

CHAPTER 8

Urban Water Harvesting

Urban centres in India are facing an ironical situation today. On one hand there is the acute water scarcity and on the other, the streets are often flooded during the monsoons. This has led to serious problems with quality and quantity of groundwater. Most of the traditional water harvesting systems in cities have been neglected and fallen into disuse, worsening the urban water scenario. One of the solutions to the urban water crisis is rainwater harvesting—capturing the runoff.

This is practiced on a large scale in cities like Chennai, Bangalore and Delhi where rainwater harvesting is a part of the state policy. Elsewhere, countries like Germany, Japan, United States, and Singapore are also adopting rainwater harvesting. The total amount of water that is received in the form of rainfall over an area is called the rainwater endowment of that area. Out of this, the amount that can be effectively harvested is called the water harvesting potential.

Potentials for Urban Rainwater Harvesting

Among the several factors that influence the rainwater harvesting potential of a site, eco-climatic conditions and the catchment characteristics are considered to be the most important.

Rainfall

i) *Quantity*: Rainfall is the most unpredictable variable in the calculation and hence, to determine the potential rainwater supply for a given catchment, reliable rainfall data are required, preferably for a period of at least10 years. Also, it would be far better to use rainfall data from the nearest station with comparable conditions.

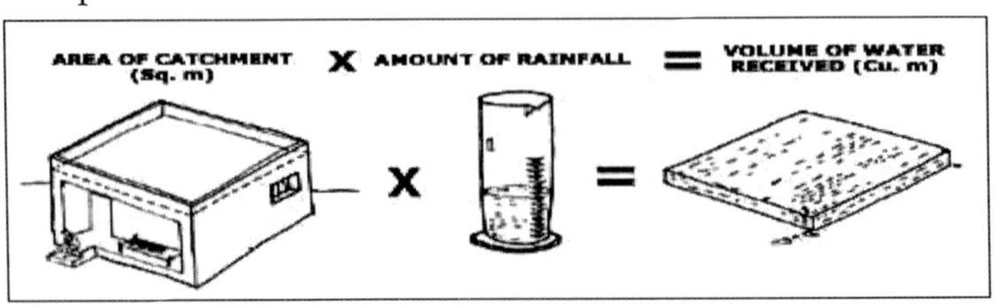

ii) *Pattern*: The number of annual rainy days also influences the need and design for rainwater harvesting. The fewer the annual rainy days or longer the dry period, the more the need for rainwater collection in a region. However, if the dry period is too long, big storage tanks would be needed to store rainwater. Hence in such regions, it is more feasible to use rainwater to recharge groundwater aquifers rather than for storage.

Catchment Area

Runoff depends upon the area and type of the catchment over which it falls as well as surface features. All calculations relating to the performance of rainwater catchment systems involve the use of runoff coefficient to account for losses due to spillage, leakage, infiltration, catchment surface wetting and evaporation, which will all contribute to reducing the amount of runoff. (Runoff coefficient for any catchment is the ratio of the volume of water that runs off a surface to the volume of rainfall that falls on the surface).

Runoff coefficients for various catchment surfaces

Type of Catchment	Coefficients
Roof Catchments	
- Tiles	0.8- 0.9
- Corrugated metal sheets	0.7- 0.9
Ground surface coverings	
- Concrete	0.6- 0.8
- Brick pavement	0.5- 0.6
Untreated ground catchments	
- Soil on slopes less than 10 per cent	0.0 - 0.3
- Rocky natural catchments	0.2 - 0.5
Untreated ground catchments	
- Soil on slopes less than 10 per cent	1.0 - 0.3
- Rocky natural catchments	0.2 - 0.5

Based on the above factors the water harvesting potential of a site could be estimated using the formula given below.

Water harvesting potential = Rainfall (mm) ´ Area of catchment ´ Runoff coefficient

Broadly rainwater can be harvested for two purposes:

i) Storing rainwater for ready use in containers above or below ground

ii) Charged into the soil for withdrawal later (groundwater recharging)

Rainwater harvesting can be harvested from the following surfaces

Rooftops: If buildings with impervious roofs are already in place, the catchment area is effectively available free of charge and they provide a supply at the point of consumption.

Paved and unpaved areas i.e., landscapes, open fields, parks, stormwater drains, roads and pavements and other open areas can be effectively used to harvest the runoff. The main advantage in using ground as collecting surface is that water can be collected from a larger area. This is particularly advantageous in areas of low rainfall.

Waterbodies: The potential of lakes, tanks and ponds to store rainwater is immense. The harvested rainwater can not only be used to meet water requirements of the city, it also recharges groundwater aquifers.

Stormwater drains: Most of the residential colonies have proper network of stormwater drains. If maintained neatly, these offer a simple and cost effective means for harvesting rainwater.

The decision whether to store or recharge water depends on the rainfall pattern and the potential to do so, in a particular region. The sub-surface geology also plays an important role in making this decision.

For example, Delhi, Rajasthan and Gujarat where the total annual rainfall occurs during 3 or 4 months, are examples of places where groundwater recharge is usually practiced. In places like Kerala, Mizoram, Tamil Nadu and Bangalore where rain falls throughout the year barring a few dry periods, one can depend on a small sized tank for storing rainwater, since the period between two spells of rain is short. Wherever sub-strata is impermeable recharging will not be feasible. Hence, it would be ideal to opt for storage.

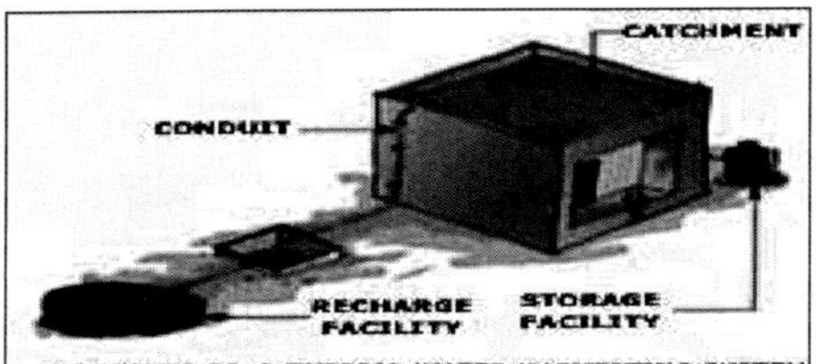

ELEMENTS OF A TYPICAL WATER HARVESTING SYSTEM

Elements of rural water harvesting system

In places where the groundwater is saline or not of potable standards, the alternate system could be that of storing rainwater.

Beyond generalisations, it is the requirement that governs the choice of water harvesting technique. For example, in Ahemadabad, which has limited number of rainy days as that of Delhi, traditional rainwater harvesting tanks, known as tankas, are used to store rainwater even today in residential areas, temples and hotels.

A rainwater harvesting system comprises components of various stages—transporting rainwater through pipes or drains, filtration, and storage in tanks for reuse or recharge. The common components of a rainwater harvesting system involved in these stages are illustrated:

i) *Catchments*: The catchment of a water harvesting system is the surface which directly receives the rainfall and provides water to the system. It can be a paved area like a terrace or courtyard of a building, or an unpaved area like a lawn or open ground. A roof made of reinforced cement concrete (RCC), galvanised iron or corrugated sheets can also be used for water harvesting.
ii) *Coarse mesh* at the roof to prevent the passage of debris
iii) *Gutters*: Channels all around the edge of a sloping roof to collect and transport rainwater to the storage tank. Gutters can be semi-circular or rectangular and could be made using:

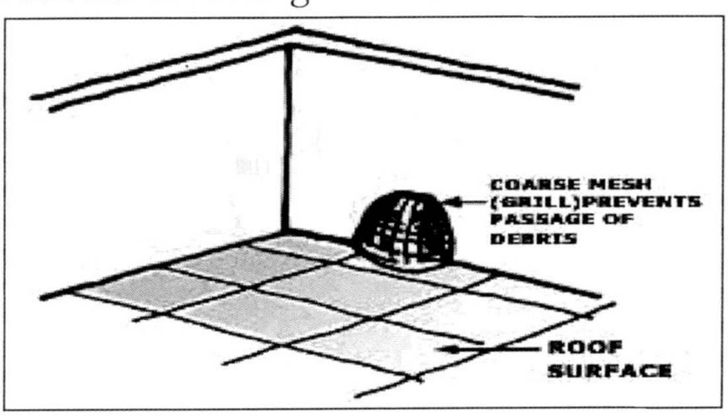

i) Locally available material such as plain galvanised iron sheet (20 to 22 gauge), folded to required shapes.

ii) Semi-circular gutters of PVC material can be readily prepared by cutting those pipes into two equal semi-circular channels.

iii) Bamboo or betel trunks cut vertically in half.

The size of the gutter should be according to the flow during the highest intensity rain. It is advisable to make them 10 to 15 per cent oversize.

Gutters need to be supported so they do not sag or fall off when loaded with water. The way in which gutters are fixed depends on the construction of the house; it is possible to fix iron or timber brackets into the walls, but for houses having wider eaves, some method of attachment to the rafters is necessary.

iv). *Conduits*: Conduits are pipelines or drains that carry rainwater from the catchment or rooftop area to the harvesting system. Conduits can be of any material like polyvinyl chloride (PVC) or galvanized iron (GI), materials that are commonly available.

v) *First-flushing*: A first flush device is a valve that ensures that runoff from the first spell of rain is flushed out and does not enter the system. This needs to be done since the first spell of rain carries a relatively larger amount of pollutants from the air and catchment surface.

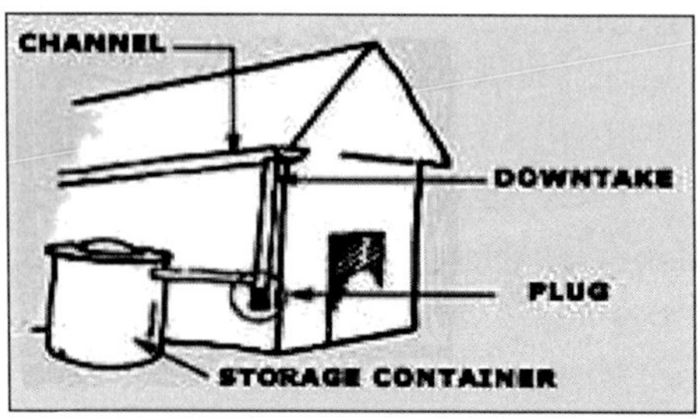

vi) *Filter*: The filter is used to remove suspended pollutants from rainwater collected over roof. A filter unit is a chamber filled with filtering media such as fibre, coarse sand and gravel layers to remove debris and dirt from water before it enters the storage tank or recharge structure. Charcoal can be added for additional filtration.

i) *Charcoal water filter*: A simple charcoal filter can be made in a drum or an earthen pot. The filter is made of gravel, sand and charcoal, all of which are easily available.

ii) *Sand filters*: Sand filters have commonly available sand as filter media. Sand filters are easy and inexpensive to construct. These filters can be employed for treatment of water to effectively remove turbidity (suspended particles like silt and clay), colour and microorganisms.

In a simple sand filter that can be constructed domestically, the top layer comprises coarse sand followed by a 5-10 mm layer of gravel followed by another 5-25 cm layer of gravel and boulders.

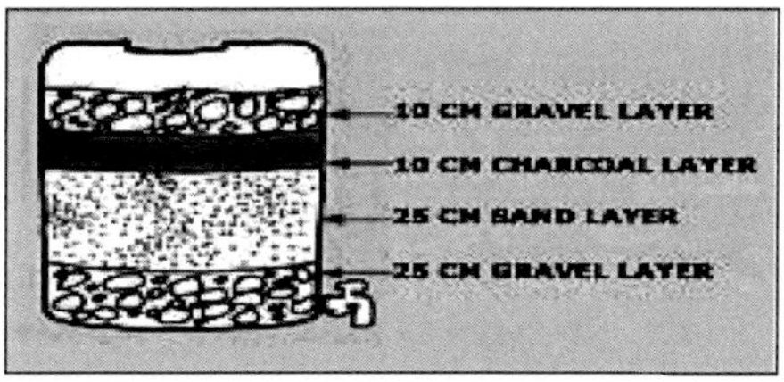

Dewas filters: Most residents in Dewas, Madhya Pradesh, have wells in their houses. Formerly, all that those wells would do was extract groundwater. But then, the district administration of Dewas initiated a groundwater recharge scheme. The rooftop water was collected and allowed to pass through a filter system called the Dewas fillter, designed by Mohan Rao , district collecter of Dewas, and engineers of the rural engineering services. The water thus filtered is put into the service tubewell.

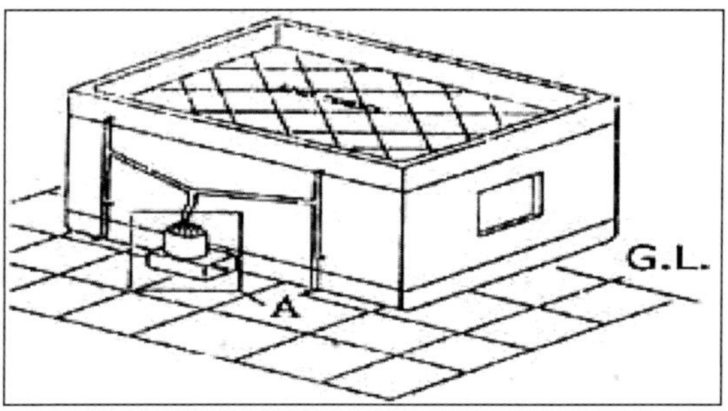

The filter consists of a polyvinyl chloride (PVC) pipe 140 mm in diameter and 1.2m long. There are three chambers. The first purification chamber has pebbles varying between 2-6 mm, the second chamber has slightly larger pebbles, between 6 and 12 mm and the third chamber has the largest - 12-20 mm pebbles. There is a mesh at the outflow side through which clean water flows out after passing through the three chambers. The cost of this filter unit is Rs 600.

Filter for large rooftops: When rainwater is harvested in a large rooftop area, the filtering system should accommodate the excess flow. A system is designed with three concentric circular chambers in which the outer chamber is filled with sand, the middle one with coarse aggregate and the inner-most layer with pebbles.

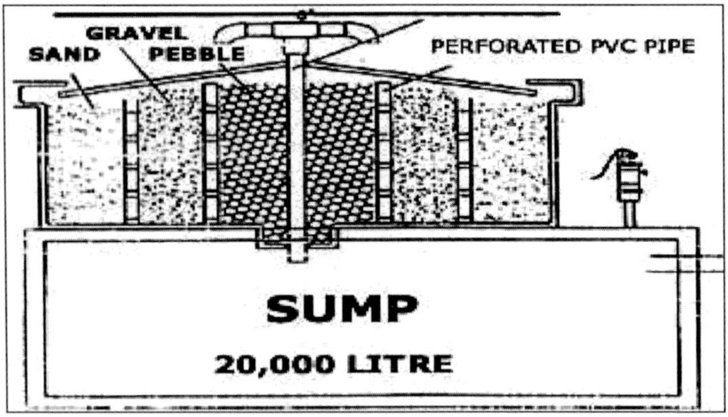

This way the area of filtration is increased for sand, in relation to coarse aggregate and pebbles. Rainwater reaches the centre core and is collected in the sump where it is treated with few tablets of chlorine and is made ready for consumption.

Horizontal roughing filter and slow sand filter: The introducton of horizontal roughing filter and slow sand filter (HRF/SSF) to treat surface water has made safe drinking water available in coastal pockets of Orissa. The major components of this filter are described below.

1) *Filter channel*: One square metre in cross-section and eight m in length, laid across the tank embankment, the filter channel consists of three uniform compartments, the first packed with broken bricks, the second with coarse sand, followed by fine sand in the third compartment. The HRF usually consists of filter material like gravel and coarse sand that successively decreases in size from 25 mm to 4 mm. The bulk of solids in the incoming water is separated by this coarse filter media or HRF. At every outlet and inlet point of the channel, fine graded mesh is implanted to prevent entry of finer materials into the sump. The length of a channel varies according to the nature of the site selected for the sump.

2) *Sump*: A storage provision to collect filtered water from the tank through the filter channel for storage and collection.

While HRF acts as a physical filter and is applied to retain solid matter, SSF is primarily a biological filter, used to kill microbes in the water. Both filter types are generally stable, making full use of the natural purification process of harvested surface water and do not require any chemicals.

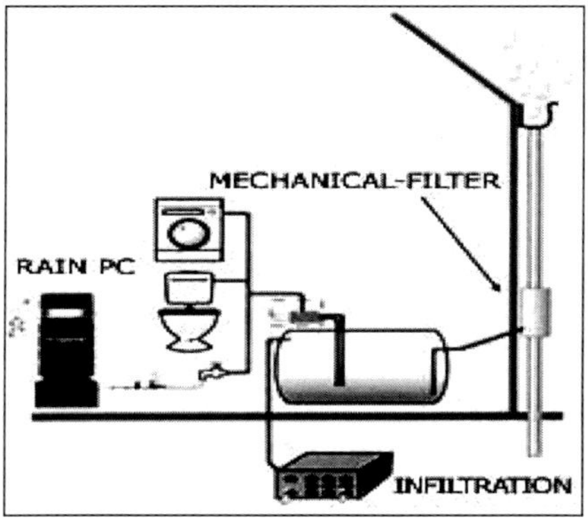

Rain PC

AcquaSure, a consortium of three specialist Netherlands-based companies, has developed a system for the conversion of rainwater to drinking water in the form of a Rainwater Purification Centre (RainPC). RainPC is developed by scaling down the multi-staged water treatment method (MST), which involves screening, flocculation sedimentation and filtration and incorporating existing technologies like upward flow fine filtration, absorption and ion exchange. Coming in a small compact 26 kg unit, the RainPC offers an affordable solution by converting rainwater into drinking water.

RainPC is made of ultra violet resistant poly-ethylene housing and cover, stainless steel rods and bolts, a nickel-brass valve and an adapter for maintaining constant volume. Xenotex-A and activated carbon catridges along with ultra membrane filtration or micro-membrane filtration modules incorporated in the RainPC has the capacity to deal with E-coli and the potential of meeting the Dutch as well as World Health Organisations (WHO) water regulation standards. The components can also be transported individually to be assembled at the site. Three product types are available based on their microbial contaminant removal capacity. This technology is ideally suited for virtually any situation and is a blessing particularly for those who have little or no access to regular safe drinking water.

The salient features of Rain PC are:

i) Simple straight-forward installation
ii) Easy to operate and maintain
iii) Needs no power and operates at low gravity pressure (0.1 bar upward).
iv) The system is capable of providing a constant flow of about 40 liters of rainwater per hour, enough for a family of five for drinking, cooking and bathing purposes.
v) Maintains nearly constant volume irrespective of water pressure.
vi) The Xenotex-A and activated carbon cartridge processes up to 20,000 liters and can be regenerated up to 10 times.

References

Bochniarz Z., "Water management problems in economics in transition", *Natural Resources Forum* 16(1): 55-64, 19923.

Davis S S., "Implementing gender policy in the water and sanitation sector", *Natural Resources Forum* 20(3): 189-198, 1996

Edwards K A, "Water, environment and development: a global agenda", *Water Resources Forum* 17(1): 59-64, 1993

Engelman R and Roy P., "Sustaining Water: Population and the future of Renewable Supplies", Washington, DC: Population and Environment Programme, Population Action International. 56 pp, 1993

Koudstal R, Rigsberman F R, Sauenije H., "Water and sustainable development", *Natural Resources Forum* 16(4): 277-290, 1992

Mathur O P., "The issue of Resources in the Context of growing urbanization", pp. 23-30, In R Krishnan (ed) *Growing Numbers and Dwindling Resources*, New Delhi: Tata Energy Research Institute, 171 pp., 1994

Roger P., "Integrated urban water resources management", *Natural Resources Forum* 17(1): 33-42, 1993

CHAPTER 9

Water Resources Management

Water Resources Management is the integrating concept for a number of water sub-sectors such as hydropower, water supply and sanitation, irrigation and drainage, and environment. An integrated water resources perspective ensures that social, economic, environmental and technical dimensions are taken into account in the management and development of water resources.

Cities are usually located near adequate or once adequate water supplies-mostly near groundwater, but sometimes near surface waters. The excessive use of groundwater has been a short-term expediency that is becoming a long-term disaster Watersheds have been allowed to become denuded, which encourages flash runoff that erodes fertile soils and results in the loss of steady water flows in the dry season. Industrial wastewater and domestic sewage have been allowed to pass untreated into major waterways, thereby endangering water supplies of downstream users. Asian cities cannot continue to go to other water catchments without first ensuring that they have optimised water availability and use within the vicinity of urban demand.

Over the last hundred years there have been two major shifts in water management. First, the people have given over their role to governments. Second, people rely less on using rainwater and have instead exploited rivers and groundwater through the use of dams and tubewells. This second fact has led to a growing and in some places unbearable stress on water from these sources. In India, the exploitation of groundwater has been encouraged, but little has been done to recharge it. As a result, groundwater tables nationwide are falling. In years when rains are low, this problem becomes an emergency.

Watershed Rehabilitation

In developing countries, especially those with large populations, the governance associated with the use of watersheds for other purposes is invariably quite weak. Furthermore, the rehabilitation of watersheds is a long-term undertaking that does not find much support when measured against the need for short-term political gains. When watersheds are denuded of trees, through logging and cutting for firewood, rainfall rapidly becomes runoff, which causes the erosion of fertile land and reduces the time during which this water can be used, unless it is stored using dams. This situation is of course exacerbated by the wet and dry seasons that dominate the climates of many Asian countries.

Watershed rehabilitation will facilitate the retention of rainwater, which will help river flows become perennial. In Sri Lanka, the National Water Supply and Drainage Board has been experimenting with watershed rehabilitation in different topographical and soil regions with a view to determining in each region the appropriate amount of integrated ground cover that would also give farmers the potential for revenue generation. It is possible to reforest more than 100 million hectares of degraded and unproductive land in India. Watershed development of 100 million hectares at the present norm would cost about $10 billion. At $1 billion per year, assuming a 10-year completion period, the cost is insignificant compared with the enormous benefits that would be derived.

In terms of water storage, flood control, reduced soil erosion and loss of nutrients, and additional agricultural crops, fuel, timber, and other produce, the value of India's reforestation in monetary terms would be much more than the total value of all the country's current industrial assets.If even a tiny fraction of the value of this great asset was spent on its upkeep, and if traditional (village) rainwater storage systems are revived, India's economy would improve more rapidly than by any other means.

Water tariffs must be raised to meet the costs of water supply, which continue to increase due to more advanced treatment, greater distances to be traveled, lower groundwater tables, and more costly distribution in densely populated areas. As tariffs rise, rainwater harvesting and other options become attractive. To collect water off the roof of a home or office and store it for future use is efficient. While there are some constraints, such as the room for storage in very densely populated areas, there is tremendous potential not yet harnessed. What is required is for local authorities to amend the bylaws so that no new home is built that does not have rainwater collection from the full roof area and storage for that water on-site.

In addition, for the owners of existing houses to also comply with these bylaws, time frames of 5-10 years can be given for the modification of the roofs and grounds of all existing homes. Chennai (India) is an example of a city making good progress in rainwater harvesting.

India receives most of its annual rainfall in just 100 hours. If this water is not captured, there will be shortages. When water harvesting and watershed development programs are handled well, it has been shown that rainwater harvesting is the starting point for meeting drinking water needs, eradicating rural poverty, generating massive rural employment, and reducing distress migration from rural areas to urban areas. Because smaller catchments give much more water than larger catchments, one state is planning to construct 10,000 check dams (one dam with a catchment of 10 hectares will collect much less water than 10 smaller dams, each with a catchment of 1 hectare). If the Government puts its mind to it, drought can be banished completely in a maximum of 10 years. The problem is one of mindsets. Rainwater harvesting demands a new approach to governance, a participatory form of governance instead of a top-down bureaucratic one.

Check Dams and Wetlands

There is no doubt that one major frustration of water resources management in Asia is the failure to capture the wet season runoff so that it can be used throughout the dry season. In the past, too much attention was given to building large dams. Now, more attention is being turned to small (check) dams. Kathmandu Valley in Nepal is an example of a location where many check dams might be used to store water during the wet season. When urban water supply tariffs are raised to around $0.40-0.50/m^3, it may become attractive for the private sector to build check dams that would divert wet season flows into Sri Lankan style "tanks," or even artificially constructed underground storage, so that the water collected can be sold to the utility for urban supply purposes in the dry season.

The dams of the estimated 200 million beavers that once lived in the continental United States made meadows out of forests, as wetlands slowly captured silt. These dams resulted in a remarkable and uniform buildup of organic material in valleys, a checkerboard of meadows throughout the woodlands, and a great deal of edge (the fruitful zone where natural communities meet). Beavers are a keystone species. Their dams create wetlands that provide homes and food for dozens of other species, including ducks, moose, fish, frogs, and great blue herons. Wetlands clarify water and prevent soil from washing downstream.

When beavers make a series of dams and ponds within a drainage basin, the water cycle in an entire watershed is affected. Wetlands are like sponges. They soak up water during storms and release it slowly in drier times. Water detained in wetlands behind beaver dams is more likely to percolate down to groundwater, raising the water table and creating springs and small streams throughout a watershed. A tenth of the total land of the United States was once beaver built wetland. Now the beavers and these wetlands are gone.

Pollution Control

Water availability can drive economic growth. And, along with a ready source of labour and ready markets or transport to markets, water availability has encouraged industries to spring up in cities. Industries, however, damage waterways through pollution, discharge of raw sewage, and disposal of garbage. Incentives must be created for the relocation of industries to industrial estates, preferably downstream of cities, where their effluents can be treated before they are discharged. Town planning rules must be transparent to the public and implemented. Industries that are water intensive should be closely monitored, not only for pollution but also for their conservation measures.

In line with this, relatively high tariffs can encourage conservation. The "polluter pays" principle can be applied in urban areas, and this can be regarded as trading for the right to restrict other uses. This more than anything is a question of governance. Others may argue that one could also consider a "beneficiary pays" principle. Perhaps there is room for both.

Water Use and Access

About 90% of all water use in Asia is for irrigation. Low efficiencies result from water losses in canals through evaporation, leakage, and pilferage. They come from irrigation methods that waste water. They come from growing crops that are heavy water users, such as paddy crops, when for the purpose of generating income other crops will do just as well on much less water. They come from the salinisation of soils caused by the use of too much water. They come from the lack of mechanisms to trade customary water rights in a free market. And they come from national policies on subsidies and food security that are often in conflict.

Continuous irrigation, without proper drainage, gradually destroys land and nearby streams and rivers through salinisation. The sun's heat evaporates irrigation water, leaving salts behind. This water also flushes salts out of mineral rich soil, leaving these to dry on the surface or dissolve in groundwater and poison plant roots. Furthermore, where rainfall is limited, crops must be carefully chosen. Sugarcane is just about the worst crop choice in terms of water needed-whereas potatoes would be much more beneficial. Subsidised irrigation is one reason for waste.

Farmers rarely pay more than one fifth of the cost of operating public irrigation schemes, let alone capital costs. In most cases, farmers can cut their water use by 10- 50%. It has been shown that an investment in irrigation efficiency is usually an investment in crop and soil productivity. Better water management often leads to increased yields, a reduction in erosion, and a reduction in the amount of fertile cropland that becomes water logged, salted, or sapped of nutrients.

Demand Management and Water Conservation

Although average domestic water consumption in European cities is only 130 litres per capita, many cities in Asia use around 200 litres per capita (for those connected to piped water). This can be traced primarily to low tariffs and a lack of public awareness concerning the importance of conserving water as a scarce resource.

When tariffs are raised in many Asian cities (Chennai, Colombo, and Manila are examples) there is no drop in demand, indicating that there is no price elasticity of demand at these levels. Standpipe water is in many cases free to users, and it is common to see water being wasted at standpipes in South Asian countries. Water is also wasted through poor plumbing in homes, leaks in distribution systems, and overstorage on a 24-hour basis, which is related to users' attempts to adjust to intermittent water supplies.

In Asia, unlike Europe, there is little attempt to conserve water through toilet or shower retrofitting. Singapore is an exception, and the PRC offers some good examples of industrial and domestic water conservation. The block tariff systems employed by Asian utilities invariably have large volumes at lifeline rates, thereby providing subsidies to all, including the rich. Little effort is going into rehabilitating watersheds to secure future supplies. It is a waste of funds to treat and distribute water if it is lost or wasted. As urban populations continue to expand, there is sometimes a need to go farther afield (more than 100 kilometers) to harness new water sources, and this is becoming very costly, not only in capital costs but also in operating (pumping) costs. Unfortunately, this fact is seldom reflected in tariffs.

Scarcity of Water

It is greatly influenced, therefore, by the integrity of watersheds, overexploitation of groundwater, cost of water transport and distribution, density of populations, and water pollution levels. What is being seen (and is clearly demonstrated by the rapid rise of bottled water use) is that the urban issue is one of scarcity of clean and potable water, which is also a reason behind rainwater harvesting becoming more and more relevant for cities. But water scarcity in Asia is not manifested across the board. It is characterised more by unequal access to water.

For example, half the population in a city might get piped water and half might rely on vended water. Another example, involving irrigation, is where farmers at the head of a system might get plenty of water and farmers at the bottom of a system might miss out altogether on water for their crops. The challenge, then, is to overcome this lack of equality. Grossly underpricing water perpetuates the illusion that it is plentiful and nothing is lost when it is wasted. Many of the world's water shortages stem from failing to value water at anything close to its true worth.

Most of the world does not treat water as a scarce resource. The foremost challenge related to water scarcity in developing countries involves inefficient water use in agriculture and urban areas and by industry. Inefficient use is linked to subsidised rural and urban water use. In line with this, irrigation water is essentially not priced. The price of water in cities does not cover the cost of delivery. And capital investment decisions in all sectors are divorced from the management of this resource. It is not uncommon for water subsidies to go disproportionately to the better off.

Industrial Use

Water utilities in cities often rely heavily on revenue from industrial users to stay financially viable. In Greater Colombo, for example, the average industrial tariff is six times the average domestic tariff. Yet industry is invariably not controlled in terms of location and type, so polluting industries, like carpet factories, are located upstream from domestic water users. There are very few industrial estates (where some collective treatment of effluents can be made). There are almost no incentives for industries to relocate to areas where they will have less effect on residential zones.

There is heavy exploitation of groundwater (normally without charge) by industry, and there is little knowledge of or control over water use by industry in terms of water needs and conservation measures. In addition, it has been noted that some major industrial users of public piped water are using armed personnel to guard illegal connections. Moreover, in Sri Lanka, the sand mining of rivers-a livelihood for many and a cheap resource for construction-is allowing seawater to move far inland, which is upsetting the water supply intakes for several towns. In short, governance is a real issue in the industrial use of water in Asian cities.

Environmental Flows

The need for minimal "environmental flows" for water left in rivers during the dry season is now beginning to be recognised. These flows are still very low and probably inadequate, but recognition is at least a start. There is a need to consider the whole ecological regime of rivers, including fishing as a livelihood; fishing as a sport; recreational needs, such as swimming, sailing, and white-water rafting; flood and fertility regimes; artificial lakes with tourism potential; garbage control; and storage of wet season flows.

Conflict of Water Users

The main users of freshwater in Asia are farmers (through irrigation), next come the urban water utilities that serve domestic, commercial, industrial, and some gardening and stock raising needs. Hydropower facilities are, strictly speaking, not consumptive users, but their activities do influence others. The environment needs a certain amount of water to be reasonably well maintained. And recreation and tourism also rely on freshwater.

By far the greatest potential for conflict among users involves irrigators and urban water utilities, as urbanisation continues to grow at rates of around 5% per annum. First comes the desire to take over the water used by the irrigators, as it is the next nearest source of water to develop. Next comes the sale of farmland for subdivision and estate construction. But there is also great conflict between farmers for water. Those at the bottom end of the irrigation network invariably miss out in times of drought. Then there is the conflict that arises when water is taken from one locality or river basin and transferred to another for irrigation or water supply. Certainly the environment is greatly altered.

Livelihoods (water wheels, irrigation, and fishing) are affected, and aesthetics are impaired. Hydropower often involves dams for storage of potential energy, and these dams severely alter the regime of rivers for other users. When industrial and domestic water users mine groundwater, they are in conflict with future users, because access will be deeper and more costly, and the risk of saline intrusion will often be ignored until the damage has been done. Pollution of waters, especially by industrial users, affects other users of the water downstream as well as groundwater users. Religious use of water is greatly impaired by such activity. Finally, there is the potential conflict of use that results from local management of water versus comprehensive water resources management on a river basin scale.

Watersheds involve the conflict between urban water users and inhabitants who live in or near watersheds and cut down trees for much needed fuel for cooking. When it comes to interbasin transfer, governments often state that water belongs to the nation, but more and more now the customary rights of the settled people are being observed. In line with this, their past and future use of water must be considered. Unfortunately, where these uses conflict, it is not easy to compare apples with apples.

Once trading water rights becomes a reality, conflict resolution by politics can be eliminated. But the impetus for trading water rights must come from domestic urban water users. When they pay tariffs of around $0.40/m^3, instead of roughly $0.05/m^3 there will be something with a realistic value to trade. But as long as domestic tariffs remain low, there will be little incentive to transfer water rights.

Chile has been a pioneer in introducing tradable water rights. In 1981, Chile's water code allocated (without charge) property rights to existing users of water. Property rights for new users were sold by auction. Property rights can be used for loan collateral and are assigned for consumptive and nonconsumptive uses. Agriculture was and remains the largest consumptive use. The city of La Serena initially planned to construct the Puclara Dam to satisfy its rapidly growing water requirements. With the introduction of a water code, many farmers in the region recognised that the water rights they held had a higher value than their water's current agricultural use.

Farmers put in place more efficient irrigation systems requiring lower water volumes and sold all or part of their water rights to the city at good prices. The transfer of the agricultural water rights to La Serena has led to the indefinite postponement of the dam's construction. The reduction in water use in the agriculture sector has helped control salinity, which was primarily caused by excessive use of water. Introducing an economic value for water rights has also had positive impacts on water utilities. For example, Chile's main water company chose to invest in the reduction of NRW rather than pay for additional water rights to meet its customers' growing demand.

Role of Agriculture in Sustainable Water Resource Management

Agriculture is rapidly emerging as the main industry that will play a key role in sustainable water resources management. In order to address the many issues surrounding water use and water consumption in agriculture a multi-barrier approach is needed and some major considerations should be directed towards influencing lifestyle changes.

About 17% of the agricultural land is irrigated producing about 40% of the food globally. This translates into 70-75% of the available freshwater in the world being used by agriculture. At the same time agriculture is rapidly emerging as the greatest non-point source contributor to pollution of water. The current demand for food is increasing rapidly and according to the World Bank food production has to increase by at least 50% over the next 20 years due to population growth, a result of the improvement in diet of the one billion people that are currently malnourished, and also due to the shifts in diet as populations become more urbanised and have more disposable income to purchase meat.

At the same time urbanisation is progressing rapidly. It is estimated that 50% of the world's population lived in urban settings in 2001 and this is estimated to increase to 70% by the year 2020. Water demand by the large mega-cities is also increasing and putting additional pressure on freshwater resources. The urban-agricultural conflicts over water are caused not only by water demand but also by the challenge of maintaining water quality.

The land base for food production is declining steadily and the availability of water for irrigation is becoming increasingly limited. In fact the expansion of irrigated land is slowing globally and in many parts of the world is actually declining due to salinisation, over-pumping of aquifers, and insufficient new water sources.

The increase in meat consumption has been particularly apparent in urban centres. While the per capita meat consumption has levelled off in North America and Europe, meat consumption in the developing world is growing at twice the rate of population growth. The demand for meat results in large increases in water demands for agriculture. In addition, the efficiency in water use in urban areas is relatively low particularly in non-water stressed countries and the irrigation efficiency in agriculture is even lower. In the urban setting most water is used for dilution of pollutants, while less than 5 % of all irrigated agriculture is done by efficient irrigation systems such as sprinklers and drip systems.

In both cases there is a great need to control demands on water resources and water conservation programs need to be initiated if we hope to prevent major water crises and conflicts in the near future. Additional efforts are needed to reduce meat consumption and this requires changes in diets that are far more challenging to achieve.

Agriculture and Water Pollution

Agriculture is the most important source of water pollution. Almost all freshwater resources (lakes, streams, and groundwater) are adversely affected by agriculture. The introduction of sediments, excess nutrients, pesticides, antibiotics, trace metals, and microbial problems associated with animal production are rapidly becoming a major problem for the aquatic biota and drinking water sources. While fertiliser application rates have levelled off in the developed world a rapid increase in its use is taking place in the Third World.

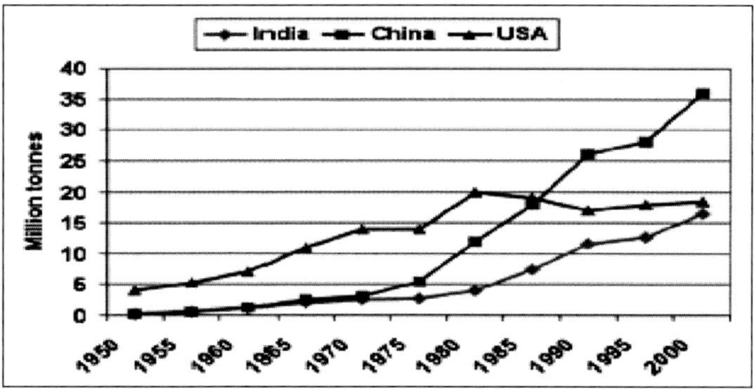

As shown in Figure 1, China has increased the use of fertiliser by 100% over the past 15 years and India had a 3 fold increase in fertiliser use. This has taken place at the same time as manure production has increased dramatically worldwide. Excess nutrient application from fertiliser and manure is leading to widespread eutrophication, which is by far the single most important problem associated with agricultural water pollution. At the same time the introduction of sediments and microbial problems associated with manure applications have created a situation where drinking water supplies are threatened.

Microbial communities often use sediment particles as a host and are able to survive for long periods in biofilms around particles. What we do not fully understand are the conditions under which these organisms flourish. Oxygen, carbon and nutrient content and temperature are likely critical factors that determine survival and outbreak of disease and a major effort is needed to gain a better scientific understanding of these factors and processes. Another critical issue is the use of antibiotics in industrialised livestock operations.

An estimated 70-80% of all antibiotics produced in the world are now used for non-therapheutic use in livestock in order to prevent the risk of infections. Most of the excess of these antibiotics ends up in water supplies and drainage systems and we do not have much scientific information on their fate or on the impact of these products on the environment.

The linkages between agriculture and urbanisation is most apparent in a watershed context. Almost every watershed has a rural and urban component and each contributes different pollutants to streams. Organic contaminants and metals are the main pollutants for urban areas while sediment and excess nutrients are the main problems originating from agriculture. In most watersheds these pollutants eventually come together creating cumulative effects that are difficult to isolate. Much of the research on endocrine disrupters is in the very early stages but there is enough evidence to raise concern.

Most of the agricultural pollution is from many dispersed, non-point sources and, while the contribution from each individual land user is small, it is the cumulative effect that is critical. To assess cumulative impact is a challenging proposition for the scientific community but even more challenging for the legal community, which is very poorly equipped to deal with these types of pollution problems because the science to measure cumulative effects is notoriously imprecise. How to prevent contamination of water resources from combined effects from agriculture and urban sources is a challenge because of the non-point source nature of the problem. Clearly a multi-barrier approach is needed to address this diffuse problem.

Water Quantity Management

Agriculture will have to give up some of its share of freshwater to satisfy the urban demand. This can be done first and foremost by improving the efficiency of irrigation. The technology for sprinkler and drip irrigation has advanced very rapidly and many low cost drip systems are now readily available in the developing world. However, not all crops can be grown effectively with drip systems. In addition an increasing portion of the cereals grown in North America is fed to animals and this is a most inefficient use of water resources.

Given the rapid increase in meat consumption in China in the past 10 years it is highly likely that China will import chicken and other meat as a water conservation measure, since new areas for cereal production are very limited and water shortages are widespread in that country. Reducing water consumption in livestock operations is not as easy as improving irrigation. Hence an emphasis on reducing meat consumption would be the best option. However, this is very much against the current global trend towards fast food and meat based food consumption in the urban environment. What is required is a substantial education process and this is a formidable challenge that cannot be achieved very easily.

Prevention of Water Pollution

A typical cow produces 6-7 times as much nitrogen as a human and if all global livestock numbers are converted into livestock unit equivalents (based on the nitrogen content in the waste) it is estimated that the waste generated by livestock animals is equivalent to about 18 billion people. We treat the human waste in most major cities to some extent but we still assume that animal manure can readily be applied to land in a non-harmful manner. This is no longer the case when dealing with large industrial type livestock operations. Hence radically different approaches are needed to deal with animal waste.

Regulating stocking densities, developing annual nutrient budgets for each farm, applying fertiliser in a more precise manner, creating manure storage for those periods of the year when manure cannot and should not be applied to the lands, and creating effective buffer zones around streams and lakes in rural areas are just some of the many options available to reduce the pollution problem from agriculture. Source control is another key issue that is critical if we hope to maintain sustainable rural environments and protect water sources. A new vision for land use management is needed. The main focus should be on pollution prevention and not on end of pipe treatment.

The cost of water treatment is increasing at an exponential rate and even with the most cost effective technology we are unable to deal effectively with excess nutrient, antibiotics and pesticide residues. Probably the biggest challenge to put all these options in place is cost. Agriculture is in no position to absorb all the cost of improved irrigation systems, reduction in stocking densities, buffer zone creation, manure storage improvement, and others.

Food prices have steadily declined and profit margins for cereal production are at an all time low. What is needed is a rethinking of how we value food, the environment, and water. For example a meat tax that would be used for improving manure management would be a very effective way for urban residents to bear some of the environmental costs of meat production. Subsidies for creating buffer zones, restructuring water pricing, and concessions to conservation farming are all options that can have very positive effects on reducing the pressure on agriculture.

References

FAO, "Water Hervesting for Improved Agricultural Production", *Proceedings of the FAO Expert Consultation*, FAO Rome, 424 pp., 1994.

Schultz, Gert A., "Use of Remote Sensing Data in a GIS Environment for Water Resources Management", *IAHS publication* 242: 3, 1997.

Tsihrintzis, V. A., Hamid, R., Fuentes, H. R., "Use of Geographic Information Systems (GIS) in Water Resources: A Review", *Water Resources Management* 10, no. 4: 251, 1996.

........., Fuentes, H. R., and Gadiputi, R. K., "GIS-Aided Modeling of Nonpoint Source Pollution Impacts on Surface and Ground Waters Water", *Resources Management* 11, no. 3: 207–218, 1997.

CHAPTER 10

Water Harvesting Due to Domestic Water Scarcity

The world population is expected to reach 8 billion by the year 2020. Food as well as water will be in high demand with increasing population. While much is debated about the worlds ability to feed the rising population, relatively less attention has been paid to population growth and household water security, specially in developing countries. Since 1970, global demand for water has increased at approximately 2.4% per annum. Rapid urbanisation and industrial development in developing countries will exert more pressure on water supply, where the majority share is allocated for irrigation. It is estimated that 1.2 billion people in 20 developing countries have no access to "Safe Water".

These developing countries have been identified as "Water Scarce", and the projections are that by year 2020 the number of countries in the water scarce category will increase to 30, of which 21 will be in low income, food deficit nations. Hence, the challenge for the future will be to ensure adequate access to safe water for the poorest households in these countries.

At a global level water withdrawals are expected to increase by 35% by the year 2020. The per capita water availability, on the other hand has declined by approximately 40% in Asia and by 50% in Africa. While large sums of funds are allocated annually to improve water supply and sanitation, number of people facing water insecurity remains above 1 billion and is on the increase.

One of the reasons for inadequate water supply coverage in most developing countries is the investment decisions in most poor countries are directed towards large-scale dam construction and/or Hydro power development, neglecting the poor communities who suffer without safe water supplies. Sri Lanka is an ideal example for this situation where 84% of the countries power supply is covered by large scale Hydro power stations and 70% of lands are irrigated with large-scale irrigation systems.

Household Water Security: A Case Study of Rural People in Sri Lanka

While this is abundant proof for availability of water resources in Sri Lanka, it has failed to provide safe drinking water to 40% of her population by 1999. Sri Lanka's per capita annual water resources stands at 2802m^3. This is expected to decline to 1928 m^3 by year 2025.

A recent study clearly indicated the high degree of seasonal and spatial variations of water supply and demand. Based on these studies Sri Lankan has been divided into absolute water scarce districts and economic water scares districts depending on the seasonality. Absolute water scarcity in this instant has been defined as when total water withdrawals are more than 50% of the available water resources. According to this classification almost all dry zone districts in Sri Lanka faces either seasonal or year round absolute water scarce conditions on the other hand economic water scarcity is defined in terms of the magnitude of future development needs.

If the future demand is less than 50% of the total available water resources, but is more than twice the level of current withdrawals, then it is considered to be in sever economic water scarce condition (Figure 1). Masking of sub-national water demand by national statistics is also apparent in the Indian context. India as a whole has adequate average water supplies of 2500 m^3 per capita, but the state of Rajastan has access to only 550 m^3 per capita of water. Similar examples have been sited from China and parts of Mexico.

In Sri Lanka, the total water demand for the three sectors, agriculture, domestic and industrial were considered for assessing water scarcity. Though no water scarcity at national level is indicated at present or in the near future, at sub-national level, large variations in water scarcity have been reported. The demand for quantity and quality of water is expected to increase in future with countries aspiration for greater degree of industrialisation. Hence, meeting increasingly competing demand will require innovative technical and organisational approaches to water resource management, backed by an effective institutional and policy environment.

Water security can be broadly defined as access by all individuals at all times to sufficient safe water for a healthy and a productive life. Household water insecurity can be simply expressed as the difference between water deficit at household and water available for household use. This can be expressed as;

Water Insecurity = Water Deficit in a Household > Available Supply at Household

Household water deficit means, the water needed over and above what is available at home in order to provide adequate water for all members of a household at all times to live a normal life. Household water deficit is a result of water consumption requirement at household level, water supply at household, distance to fetch water and availability of water carriers. Each of these factors can effect the household water security either positively or negatively. However, impact of these factors on household water security depends on household size, number of growing children, type of work done by householders, health of water carriers, old people at household etc.

Among rural communities in Sri Lanka, the socio-economic statuses of the sample householders vary from casual daily wage earners to small-scale tea landholders. However majority are small-scale farmers by profession. Monthly income of the sample population varies from less than Rs. 1500 to Rs. 10,000 per month depending on the profession of the householders. The three agroecological zones were selected to assess the water use pattern of households living primarily in different rainfall regimes and to understand the multiple water source use by rural communities.

In the flat terrain dry zone, it would be the distance to fetch water but in the wet zone in southern Sri Lanka the hill-terrain makes travelling much difficult with water containers than doing the same in the dry zone. In the intermediate zone, the sample was located in a manner that had a mixture of long distances and hill-terrain to climb. The comparative water use data and behavioural information from wet zone tea small holders, daily wage earners from the intermediate zone and small-scale farmers from the dry zone would give us the relative importance of direct costs and opportunity costs of fetching water under different economic conditions.

Among rural communities in Sri Lanka water quality perceptions have been a strong deterrent to drinking water security. Most households do not drink rainwater not because of the absolute poor quality of water but because of the perception that rainwater harvested from rooftops is inherently poor. Other reasons contributing to this situation are the absence of filters and first flush systems, contaminated roof run off with organic pollutants like leaf mater and bird droppings.

Water for Domestic Use

At present there are dug wells used separately for drinking and bathing purposes. These wells are located at the valley bottom where water transport becomes extremely difficult and during the dry season water transport time becomes a significant factor in household time budget. Roof Water Harvesting is of two types a) Using temporary gutters to collect rainwater into open rectangular tanks. b) Collecting roof water to Brick dome/Ferro-cement tanks of $5m^3$ capacity constructed under the community water supply sanitation project (CWSSP). This system uses more permanent gutters and the water tanks are more robust and have a life span of 10 years.

Uses of Roof Water

The rainwater collected in small rectangular tanks is open at the top, hence subjected to contamination. Maintenance is minimum in these tanks, thus, water can be used for about week after each rainfall. Householders use this water only for personnel washing and toilet use immediately after rains. Tanks constructed under the CWSSP are more durable, covered from the top, hence water can be used for a long period of time. Householders use this water for all household purposes, including drinking (occasionally) and cooking.

Drinking Water Security

Drinking water wells and bathing water wells. People use a strict criteria in selecting drinking water wells.

i) Drinking water wells should not be used for bathing
ii) Surroundings of drinking water wells should be cleaned
iii) These well should always have flowing water
iv) It should be used by number of families
v) Distance to the drinking water wells
vi) Ownership of the well

However during the dry season water level in most of these wells deplete making water fetching difficult. Another common problem is lack of able labour to fetch water from these wells. Usually male children and adult males are involved in fetching water. If they have to leave the house for few days they fetch adequate to compensate for their absence. When water carriers fall sick or the only male in the family is old, the household faces a situation of water insecurity.

Rainwater Harvesting beneficiaries, use rain water after boiling to over come the drinking water insecurity period. However, they use rainwater directly for drinking during the rainy season. This ensures a better quality of water during continuous rain. Apart from roof water harvesting few households use directly harvested rainfall for drinking. However, only small quantities of rain water can be harvested in this manner and only small vessels found in the kitchen are used for this purpose. Thus, harvesting rain water has increased the drinking water security of these peasants.

Cooking Water Security

Rainwater tank beneficiaries use stored water for cooking, especially during rainfall periods. However, few days after cessation of rainfall, people discontinue using rainwater due to change in colour of water, smell and sometimes due to mosquito larvae. However, if good quality well water can not be found due to non-availability or due to lack of labour, people restore to rainwater even if the quality is not premium. In such situations, secondary quality of stored rainwater is used for washing vegetables and rice before using premium quality well water for actual cooking.

Toilet Water Security

Water for toilet use is predominantly stored rainwater. When the rainwater tank water level recedes, people bring a bucket of water for toilet use whenever they visit a well for any other purpose. However, when supplying water is extremely difficult, they use pit toilet instead of water-seal type toilets.

Bathing Water Security

Specially identified dug wells are used for bating. These wells do not have a parapet wall and fertiliser residues from cinnamon plantations often contaminate water. Wells are sometimes located in their own compound, in such event the distance they have travel vary from 100-150 meters. However, when they do not have wells in the compound, they could be travelling as much as 1½ km to bathing wells. However, when there is adequate rain, specially women bath with stored rainwater or when householders have to bathe in the night after days work, they often restore to use stored water. In both situation these people have improved their bathing water security due to the presence of stored rainwater.

Water Security for Washing Cloths

There is no prioritisation of water for washing cloths. However, people do not use well water for washing cloths if the water is contaminated and the colour of water is muddy or brown. In such situations they use wells with clear water.

Water Sources for Domestic Use

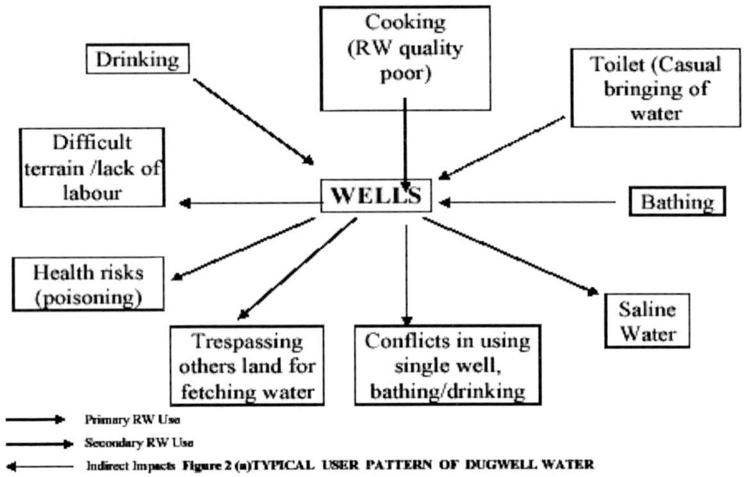

Figure 1. Typical user pattern of dugwell water

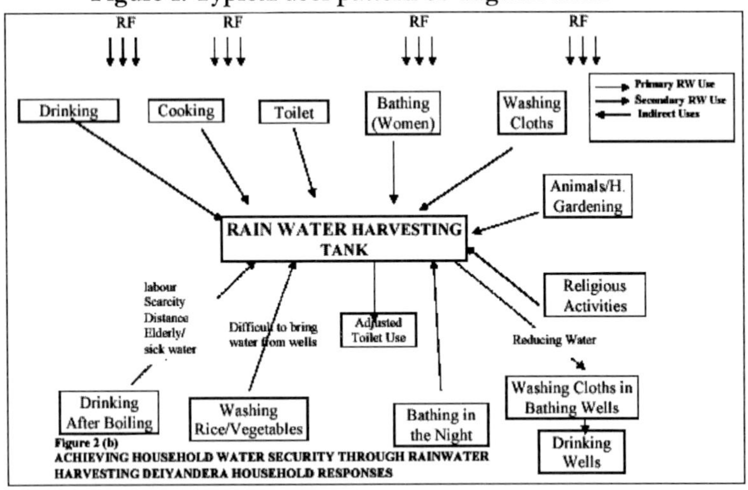

Figure 2. Achieving household water security through rainwater harvesting

There are four types of water sources used by rural people. The first source of water is stored rainwater collected during rain, secondly well water from outside the village where water is not saline and drinkable. Thirdly town water supply where wealthy people fetch only for drinking. Finally the saline water wells found in the village which people use under difficult conditions.

Rainwater Harvesting is of three types, collecting rainwater into small vessels during rainfall. Collection under this system can be used for short duration and collected quantities are small. There is no system of storage under this method. Harvesting roof water into more permanently built brick-motor tanks. This system collects more water for future use though water is subject to contamination. The third method is harvesting rainwater from tree-trunks, which incidentally is a traditional method practiced in Sri Lanka, under water short situations.

Quality of Drinking Water Security

Quality of any water directly determines the drinking water security of a community. In the two study locations too, it was the physical quality of water that decided whether the water was good for drinking or not. This theory applies to both stored roof water and dug well water. In this respect, it was easy to assess the physical quality of water in a roof water tank than in a dug well. Thus the latter has a defacto advantage over rain water. The criteria generally used by these communities to assess the quality of water are as follows;

1. Colour of water
2. Taste of water
3. Presence of mosquito and other types of larvae
4. Odour of water

The community perceptions of drinking water are governed by these factors. Besides these physical characters of water, selection of dug wells for drinking water also depends on the depth of wells, flowing water in wells and number of users per well. People always prefer to use wells used by a larger number of families rather than few families. This implies that larger the

number of users better the quality of water.

According to community perceptions, lack of filters and non-adherence to first flush of water makes stored water unfit for drinking. As a result of this situation, decaying leaf litter and other organic derbies are common in most rainwater harvesting tanks. People also judge the quality of water depending on the level of water in the tank or period of continuous stagnation without any replenishment from rainfall. People in the two study locations firmly believe that water begins to deteriorate when rain water tank water level is down to ¼ of the tank. This implies that continuous stagnation of water over a long period encourages water quality deterioration.

While the same community would use stored rainwater even for drinking, during and immediately after rains. This implies that people believe that quality of rainwater is good for drinking when it is fresh but contaminates over time. This perception of people do not change whether the storage tank is underground or above ground. Hence, the drinking water security of rural households depend on intensity and duration of rainfall with respect to stored rainwater.

To substantiate the water quality perceptions of the community, water samples from underground rainwater tanks were tested for its physical and biological quality. When water sample were tested on 26th June after month of no-rain period, 6 out of 8 tanks had good quality water with no indication of E-coli. The two tanks with no tank covers recorded an E-coli count of 100/100 ml. In one of these tanks, water was extracted using a bucket, which could be a source of contamination. When the same sample was tested on 18th August after rains, all tanks except one indicated the presence of E-coli. In the first instance the tanks were half filled with water while in the second instance the tanks were less than quarter filled with water. The change in quality of water can be attributed two reasons.

1. Since non of the tanks had a first flush system washing of birds and possibly rat and Bandicoot/Polecat droppings would have contributed to the presence of E-coli in water samples immediately after the rains.
2. When water is stored for a long time (as in the earlier case) with out any new replenishments, absence of sunlight and minerals can impede bacterial growth and also could lead to elimination of existing bacteria. This could have been the reason for non existence of E-coli in the water samples taken on 26th June.

Modelling of Water Security

Metered supply is provided to households while free standpipes and bousers are available for low income urban communities. Urban communities experience water cuts during relatively short dry weather periods due to low flows in major tributaries. However, they have shown resilience to such situations by obtaining water from other sources and by reduced water usage. Urbanisation has taken place in cities located mainly in the wet and intermediate zones, with significant rainfall distributed throughout the year. Though there exists a high potential to use harvested rainwater, provision of high quality water at low cost has discouraged initiating such endeavours.

The peri-urban communities largely depend on groundwater, obtained through large diameter wells. Households either have their own wells or have access to a private/or a public well. Wells identified exclusively for drinking provide good quality water and are protected from contamination. The amount of water available for domestic use cannot be defined in a conventional manner since a higher percentage is re-used after percolation. In such instances, a high water security is maintained while maintaining a high consumption level.

Rural communities mainly rely on surface and ground water bodies to meet their domestic requirements. Their use includes water for animals and home gardening. During extended dry weather periods, quality water is not available nearby, and has to be obtained from distant sources. This has resulted in more time being spent on collecting, transporting and storing water, reducing the overall productivity of a household.

Household Water Security should reflect the community's ability to access required quantity and quality of water for consumption, hygiene, other household needs and for minor economic activities. Water available may have its short term variations due to hydrological changes as well as long term variations due to hydrological changes as well as long term effects, resulting from socio-economic stresses, affecting households, communities, a country, or a region. Household Water Security can be expressed in terms of water availability (WA) and overall water use, $\sum_{1}^{n} W_i$ (in litres per capita per day (lpcd)), as:

$$WS_{1,2,,n} = \frac{WA}{\sum_{1}^{n} W_i} \quad (1)$$

where water requirement can be identified as drinking, W_1, cooking, W_2, washing, W_3, toilet needs, W_4, bathing, W_5, and others, W_6. The Overall Water Security defined in equation (1) can also be evaluated by combining water security values obtained for individual attributes such as: for drinking:

$$WS_1 = \frac{WA_1}{W_1}$$

and for cooking:

$$WS_2 = \frac{WA_2}{W_2}$$

where the Overall Water Security Index can be 'weighted' to reflect the relative importance.

The complexities in water use arising due to human perceptions (such as taste, cultural values) cannot be easily quantified. Models based on Fuzzy Set Theory can be used to express both qualitative and quantitative data and the results obtained for individual water uses can be integrated to give an Overall Water Security Index.

Fuzzy Set Theory

Fuzzy mathematics was first introduced by Lotfi A. Zadeh and is extensively used in analysing non-deterministic problems such as in the evaluation of slope failure potential. Fuzzy models possess a higher degree of flexibility to model complex field situations. It has the capability of representing the input parameters in terms of qualitative ratings expressed in linguistic terms. Fuzzy mathematics uses Fuzzy Sets to model the relative values of attributes and their relative importance expressed as fuzzy 'ratings' and 'weights'. They are qualitatively expressed in linguistic terms characterised by fuzzy functions normalised to the domain ranging between 0 to 1 (Figure 3).

Decision Tree Analysis

A 'Decision Tree' represents the hierarchical structure which establishes the overall attribute; in this case the Overall Water Security Index. The proposed model uses a decision tree consisting of a single level, i.e. primary factors and their respective weights, which are integrated to yield the Overall Water Security Index. It is assumed that the factors at these levels have an independent contribution towards the final result.

References

Ariyabandu R.de.S, and Dharmalingam S., *Harvesting Rainwater: A means of Water Security to Rural Sri Lanka*, HARTI, Cololombo, Sri Lanka, 1997.

Clarke, R., *Water The International Crisis*, Cambridge, Ma: MIT Press, 1993.

Gleick. P.H., "Basic Water Requirements of Human Activities: Meeting Basic needs", *Water International* 21, 83 - 92, 1996.

Webb. P. Iskandarani M., "Water Insecurity and the Poor", Issues and Research Needs", ZEF Bonn Center for Development Research University of Bonn, 1998.

CHAPTER 11

Watershed Management

A watershed is an area that drains to a common point. It may be managed for various objectives, depending on local needs, including capturing runoff, minimising erosion, and reducing nonpoint source pollution. In management of small watersheds, capturing runoff for local use is conceptually equivalent to harvesting water. Water-harvesting systems either concentrate water into a storage reservoir or apply water directly to the soil in the cropped area. Both types of systems can vary in scale from a few square meters benefiting a single household to a few square kilometers serving a larger group of people.

Storage Reservoir

Systems that concentrate water into storage reservoirs can be used for a variety of purposes, including household, irrigation, or livestock consumption. Rooftop water-catchment systems provide domestic water in many places, especially dry areas with inadequate municipal supplies. A rooftop that is 50 square meters can supply an annual average of 50 litres per day with 500 millimetres annual rainfall. This supply can help reduce competition between agricultural and household demands and can free women from the chore of collecting water.

Harvesting water to collect it in village ponds is an ancient system in some places, including southern India. The principle behind this approach is to concentrate water spatially and temporally. In arid and semi-arid areas where rainfall is low and variable, harvesting rainwater allows users to conserve it until enough water is collected to reliably support a crop. The volume of water stored at the end of the rainy season determines how much area can be cultivated.

Traditional water-harvesting systems store water in ponds and reservoirs and deliver it by gravity. With the spread of motorised pumps in recent decades, water can now be captured in some places using the same techniques but allowing collected water to percolate into groundwater aquifers. The systems are small in scale, drawing from gullies and microcatchments to recharge groundwater that supplies a number of local wells.

Soil-Moisture

In many traditional water-harvesting systems, runoff water is channelled directly to the cropped area during rainfall and stored in the soil. Where rainfall is unevenly distributed and soils have high water-holding capacity, this system may store water until the end of the rainy season, when a crop is grown under gradually receding moisture. Where soils are sandier and do not retain moisture for a long time, moisture may be channelled spatially to the location where crops or trees can take advantage of it. Farmers in West Africa commonly use such systems, cultivating their dryland crops behind a variety of small earthen barriers designed to capture moisture. These systems can remain productive several months into the dry season when the surrounding land is barren.

Water-harvesting systems vary a great deal in the ratio of catchment to cultivated area. In dry areas with less available water, this ratio can easily reach 20:1. In such water-harvesting systems, cultivation systems must be extensive and located in sparsely populated areas so that command and catchment areas do not interfere with each other. Where rainfall is higher, the ratio declines-sometimes to the point that the entire catchment area lies within a single plot.

Water-harvesting Constraints

Constraints to developing improved water-harvesting systems and adapting existing ones for productive use relate to technical feasibility, financial viability, and social organisation.

Technical Feasibility

The key technical question for agricultural water-harvesting systems is whether capturing and storing enough water to raise crop yields is feasible. In field conditions, technical feasibility depends in large part on local agroclimatic conditions. For example, in India the Indo-German Watershed Development Project works only in villages where the terrain is conducive to water harvesting -that is, there is a sloped, uncultivated catchment area and a good opportunity to build storage ponds.

Financial Viability

Even where water harvesting is technically feasible, it may not be cost-effective. A typical problem in storing water is the need to line catchment or storage areas so that water can be stored until dry periods when it is actually needed. However, using a plastic or concrete lining can raise costs beyond the financial returns. An example of where lining is most likely to be cost-effective is in mountain regions where water enables farmers to grow counterseasonal crops that can be sold in the plains at high prices.

The main constraints to rooftop collection systems are the need for a tile or sheet metal roof that can capture runoff and a storage tank, both of which can be prohibitively expensive in very poor areas. In that case, village-level systems housed on public buildings may be more viable, along with smaller-scale household systems. The key issue with storage tanks is how to make them large enough to store plenty of water while also keeping costs down. Given growing water scarcity and the high financial, social, and environmental costs of alternatives, governments may find subsidising water-harvesting systems attractive, especially those for domestic supply. Some countries already subsidise rooftop collection systems.

Social Organisation

Water harvesting requires collective action in densely populated areas where more than one person uses catchment and command areas for multiple purposes. The key question in such cases is whether the benefits of water harvesting exceed the costs of coordination among users. Often this can be the most daunting constraint of all, especially where some resource uses are mutually incompatible and any intervention will impair at least one potential use, as in India.

Successful water harvesting requires protecting upper reaches of small catchments against erosion that would reduce water-storage capacity in the lower reaches. Typically, upper catchments are denuded, so protecting them requires limiting their use for grasing and fuel collection. This restriction imposes the greatest costs on landless people who depend the most on these areas and who do not stand to gain directly from water harvesting. Long-term success requires devising institutional approaches to ensure that landless people gain from the water-harvesting intervention.

References

Bhaduri, B., Harbor, J., Engel, B., and Grove, M., "Assessing Watershed-Scale, Long-Term Hydrologic Impacts of Land-Use Change Using a GIS-NPS Model", *Environmental Management* 26, no. 6: 643, 2000.

Chang, T., Hsu, M., Teng, W., and Huang, C., "A GIS-Assisted Distributed Watershed Model for Simulating Flooding and Inundation", *Jounal of American Water Resources Association* Vol 36, No. 5: 975, 2000.

Schreier, H. et. al., "Environmental Auditing—Gaining Forests But Losing Ground: A GIS Evaluation in a Himalayan Watershed", *Environmental Management* 18, no. 1,: 139–150, 1994.

CHAPTER 12

Socio-economic Factors of Water Harvesting

Widespread adoption of water harvesting techniques by the local population is the only way that significant areas of land can be treated at a reasonable cost on a sustainable basis. It is therefore important that the systems proposed are simple enough for the people to implement and to maintain. To encourage adoption, apart from incentives in the form of tools for example, there is a need for motivational campaigns, demonstrations, training and extension work.

It is tempting to assume that a system which works in one area will also work in another, superficially similar, zone. However there may be technical dissimilarities such as availability of stone or intensity of rainfall, and distinct socio-economic differences also. For example a system which is best adapted to hand construction may not be attractive to people who normally till with animals. If a system depends on a crop well accepted in one area - sorghum for example - this may be a barrier to acceptance where maize is the preferred food grain.

The objective of rainwater harvesting projects is to assist resource-poor farmers to improve their production systems, it is important that the farmer's/agropastoralist's priorities are being fulfilled, at least in part. Otherwise success is unlikely. If the local priority is drinking water supply, for example, the response to water harvesting systems for crop production will be poor.

Peoples' Participation

It is becoming more widely accepted that unless people are actively involved in the development projects which are aimed to help them, the projects are doomed to failure. It is important that the beneficiaries participate in every stage of the project. When the project is being planned, the people should be consulted, and their priorities and needs assessed. During the construction phase the people again should be involved -supplying labour but also helping with field layouts after being trained with simple surveying instruments.

Throughout the course of the season it is helpful to involve people in monitoring, such as rainfall and runoff and recording tree mortality. A further participatory role is in maintenance, which should not be supported by incentives. After the first season it is the farmers themselves who will often have the best ideas of modifications that could be made to the systems. In this way they are involved in evaluation, and in the evolution of the water harvesting systems.

Gender Equity

If water harvesting is intended to improve the lot of farmers in the poorer, drier areas, it is important to consider the possible effects on gender and equity. In other words, will the introduction of water harvesting be particularly advantageous to one group of people, and exclude others? Perhaps water harvesting will give undue help to one sex, or to the relatively richer landowners in some situations. These are points a projects should bear in mind during the design stage. There is little point in providing assistance which only benefits the relatively wealthier groups.

Land tenure Issues

Land tenure issues can have a variety of influences on water harvesting projects. On one hand it may be that lack of tenure means that people are reluctant to invest in water harvesting structures on land which they do not formally own. Where land ownership and rights of use are complex it may be difficult to persuade the cultivator to improve land that someone else may use later. On the other hand there are examples of situations where the opposite is the case - in some areas farmers like to construct bunds because it implies a more definite right of ownership.

Management of Land

The whole question of land management by village communities has recently been acknowledged to be extremely important. Degraded land in and around villages can only be improved if land use management issues are faced by the communities themselves. One of the techniques which can assist in rehabilitation of degraded land is water harvesting - but it is only one tool among several others and cannot be effective in isolation. Unless, for example, grazing controls are implemented, there is little point spending money on water harvesting structures for re-seeding.

Project Related Water Harvesting

The experience of projects related to water harvesting and soil conservation has shown that there is no substitute for dialogue with the farmers/villagers, and a continued close relationship throughout. Projects should always aim to learn from the people of the target area, in particular about local traditional technology. It is essential that project authorities keep in mind the importance of people's priorities and participation. It is important that the benefits of the new systems should be apparent to the farmer as early as possible.

For new techniques there is often a need for demonstration before people will understand and envisage their effectiveness. Motivation and promotion of awareness among the people with regard to the project objectives and how to achieve them are very important issues. There are two basically different approaches with regard to water harvesting projects.

Demonstration, Training and Extension Approach

The technology introduced by the project is relatively simple, and costs per hectare low. The intention is to promote systems which can be taken up and implemented by the people themselves, with a minimum of support. The philosophy behind this approach is that the people themselves must be the prime movers in the development of their own fields and local environment.

Implementational Approach

In this approach the technology may be simple or complex, but it is implemented by the project itself. Machinery is often used, but some projects employ paid (or otherwise rewarded) labour. Costs are often relatively high. The intention is that the project will quickly and efficiently rehabilitate land for the people. The philosophy is that the people are simply unable to undertake the extent of work required using their own resources and therefore they require considerable or complete support to implement the project.

Experience shows that it is the first approach which offers the most hope for sustainability once the project has come to an end. Nevertheless there are situations where the introduction of appropriate machinery or support of some labour can be justified.

Machinery for Conservation

This question has been touched on already - but it is an extremely important issue. The introduction of inappropriate heavy machinery for conservation structures has been a mistake repeated widely over Africa. Conversely some mechanisation—especially where animal traction is a component—can immeasurably speed up work rates and reduce drudgery. The advantage of working by hand is that the people regard the techniques as within their capability. As long as part of the work is voluntary, they will be more willing to carry out maintenance. Nevertheless hand labour is slow, and labour shortages can be a serious constraint in some areas.

Flexibility of Approach

Water harvesting and conservation projects should never have fixed work plans or rigid targets, at least not in the early stages of implementation. The reason quite simply is that it is unrealistic to plan for all contingencies, and arrogant to assume that the techniques and approaches planned from the outset cannot be improved Learning from experience, and from interaction with the people, is a much better approach. Flexibility should be written into every project document.

Subsidies and incentives

Many water harvesting projects provide subsidies or incentives for construction. Several points need to be made about these:

i) help and assistance should only be considered as stimuli to the programme; too big a subsidy to begin with can cripple future expansion and deter participation.

ii) it is important that in all cases the beneficiaries should make at least some voluntary contribution towards construction. The level of contribution should rise when incentives are provided.

iii) food-for-work is common in projects in drought-prone areas. It is not easy to manage food distribution and development work at the same time. Generally other incentives, such as tools for work, are preferable.

iv) incentives/subsidies should not be used for maintenance: this should be the responsibility of the beneficiaries.

Monitoring and reporting

Monitoring, evaluation and reporting are often weak spots in water harvesting projects. Too many projects fail to collect data at even the most basic level. For example crop yields and tree heights are often just estimated. It is also very rare to find any information on the frequency or depth of water harvested. Without a basic monitoring system, projects are starving themselves of data for evaluation. Without clearly written reports, widely circulated, projects are denying to provide others with important information.

Water Pricing Policies

Water scarcity depends on location, climate, season, and potential use by humans. The scarcity value of water is determined by the quality of water, the quantity of water, and the number and type of water users. Water pricing policy has the potential to mitigate water scarcity. Because of its key role in managing water demand and augmenting water supply, water pricing is an important policy instrument for creating incentives to conserve and allocate water efficiently.

By providing financial justification for developing additional supplies from conventional and unconventional sources, pricing policy can make more water available to users. Unfortunately, the water pricing policies being pursued in most countries fail to perform these vital roles due to faulty approaches and inappropriate institutions, both of which have their roots in political economy.

The state provides many irrigation and domestic water supply systems at subsidized rates because doing so delivers public goods in the form of secure food supplies, public health, or legitimacy for the government. But low water charges and poor cost recovery lead to declining funds available for investment in water infrastructure, poor maintenance of existing systems, inefficient water allocation, and growing conflicts between those with and without access to water. Economic and political pressures stemming from these problems have promoted a renewed debate on water pricing policy.

Financially, water pricing is the main mechanism for cost recovery. Economically, it signals the scarcity value and opportunity cost of water and guides allocation decisions within and across water subsectors. The financial function requires water rates to cover the cost of supplying water to users. The supply cost is usually calculated by adding the operation and maintenance costs and the capital costs of constructing the system. But full cost recovery also requires water rates to reflect the long-term marginal cost (the cost of supplying an additional unit of water including the social cost of externalities).

The economic and allocative role of water pricing requires water rates to capture the scarcity value (or the marginal productivity/ utility) and to equalise the opportunity costs of the resource across uses. As water moves from least productive to most productive uses, places, and time points for efficient allocation, there will be a convergence of the scarcity value, opportunity cost, and longterm marginal cost of the resource. Unfortunately, such a convergence is rarely seen in practice. Unrealised opportunities still exist and water rates can be designed to capture at least a part of these opportunity costs. For this to occur, technology to store, transport, and deliver water is required, as are institutions to govern the development, allocation, and utilisation of the resource.

Economic Incentives

The effectiveness of the financial and economic roles of water pricing policy depends on pricing methods, sectors, and countries. Water pricing may be either volumetric (based on the quantity of water used) or flat rate (based on area irrigated or households benefited). Volumetric pricing is conducive to creating incentives for efficient allocation and use, but the cost of establishing volumetric water delivery structures is often prohibitive, especially in large and spatially spread surface irrigation systems serving many smallholders. As a result, area-based fixed rates are dominant in most irrigation systems.

However, volumetric water rates are widely used in many urban water supply systems. Water rates are generally higher in countries with severe water scarcity and institutionally developed water economies. The industrial and power sectors within a country usually pay the highest water rates and receive a higher, more costly level of service throughout the year, as do domestic users. Agriculture pays the least, but also receives the lowest level of service. Within each sector and country, there can also be a wide variation in costs as domestic users buying water from private vendors pay much more than those connected to municipal systems and farmers receiving irrigation from public canals pay far less than those receiving irrigation from private wells.

Full cost recovery is legitimate both in economic and equity terms only if the users gain all the benefits from investment. On this count, some argue that since food prices fall when irrigation increases food production, consumers benefit, and that higher water rates for full cost recovery could adversely affect farmers' incomes. But this need not be the case because farmers benefit from irrigation partly from higher output and partly from increased land prices. In any case, the kind of full cost recovery being contemplated in many contexts does not cover even 10 percent of the additional income from irrigation, let alone the land value appreciation.

Water rates are still subsidised even in countries with a relatively mature water economy such as Australia, Israel, and the United States. This is rooted in the political economy of water, as powerful state and user interests often oppose charging the full cost of water. As a result, the gap is vast between the observed water rates and the ideal economic prices of water, as reflected by its scarcity value and opportunity cost. This illustrates the real magnitude of the task of designing pricing policies effective enough to play, simultaneously, the financial and economic roles.

Institutional Dimension

There is growing recognition that a realistic water pricing policy that ensures full cost recovery can be politically feasible only when it is designed to perform well in its economic and allocative roles. Also notable is the growing realisation that institutional reforms to enhance the effectiveness of water pricing policy are indispensable. Pricing reform does not end with raising water rates. It also involves concurrent changes in pricing methods and approaches and the creation and strengthening of supportive institutions.

Volumetric pricing, though necessary, is not sufficient to ensure an effective economic role for water pricing. As long as the rates are below both supply cost and water productivity and there is no upper limit on individual water withdrawals, volumetric pricing *per se* can neither achieve full cost recovery nor solve the key incentive problem. For this purpose, volumetric pricing needs to be accompanied by price levels that are consistent with prevailing economic and resource realities. A system of transferable water quotas specifying upper limits for individual water withdrawals needs to be established for water markets to emerge that can, in turn, provide a basis for determining economically consistent water rates.

Pricing reform can be effective and practical only with the necessary institutional and technical conditions that enable cost recovery and allocative roles for water pricing policy. These institutions include an independent water pricing agency and regulatory body, financially autonomous agencies to supply water, clearly defined water rights, and transfers of management to user organisations or the private sector. Technical conditions include volumetric delivery, measurement structures, and infrastructure to move water over space, type of use, and time. These institutional and technical conditions can encourage the emergence of direct allocation mechanisms such as water markets and implement water pricing reforms.

Water Markets

Water markets are more flexible than command-and-control instruments in moving water to higher-valued uses in a manner agreeable to all parties, thus promoting economic growth and diminishing social tension from competition for scarce water resources. Sometimes water markets are also expected to mitigate environmentally detrimental effects from overexploitation of water resources. However, they can also have drawbacks, and decisions about the introduction of a water market in a specific setting need to be informed by a thorough institutional analysis of the local socioeconomic, cultural, and hydrological context.

However, there are differences between water markets, and design will depend on the prevailing hydrological regime, including whether trades involve surface water, groundwater, or both; the previous existence of informal water rights and trading; the types and numbers of water users and right holders, including whether all are irrigation farmers or from different water-use sectors; and the physical arrangements for moving water between users. However, appropriate institutional arrangements must be in place to ensure positive outcomes from water markets.

Institutional Arrangements

The implementation of water markets requires both government involvement and active water-user participation. Institutional ingredients needed for other approaches to improving water-resources management-such as an administrative system that registers and enforces timely water deliveries, a transparent and accepted water measurement system, and a well-maintained delivery system-are all required for a functioning market. Therefore, the difference between the implementation of administrative systems and a market system may not be as large as expected. The principal difference is the need for the following:

i) Definable and transferable water rights. Reallocation by trading means getting compensated. That is different from administrative reallocation, where right holders may not receive any compensation when water is reallocated to a different use. Under water markets, right holders will not only consider what the water can produce for themselves, but also the opportunity cost of the water (such as the value added by using the water in car manufacturing). Thus, the highest value of water use is taken into account, providing an incentive for more efficient use and reallocation of water to a higher-valued use.

ii) *Internal institutional market arrangements.* Buyers and sellers need to find each other and have confidence in the characteristics of the goods they want to exchange. In a well-functioning water market, this implies measuring the water available for trading, developing a simple mechanism to find prospective buyers and sellers, and carrying out the transfer. If the official transfer is difficult, then informal trading will likely occur anyway. The disadvantage of informal trading is that water use is hard to track.

iii) A physical delivery system. Traded water is likely to move from one user to another. Depending on the hydrological conditions or previously constructed infrastructure, the cost of adequate infrastructure may be too high and outweigh the benefits from transfers, rendering the water price so high that no buyers can be found. The government may absorb the cost, but thorough analyses are needed to establish whether the introduction of a tradable system will produce sufficient societal benefits to warrant large investments.

The most challenging institutional feature of a water market may thus be the definition and registration of water rights, together with the implementation of adequate information and transaction mechanisms that provide equal possibilities for all water users to participate in the market. These mechanisms would have to include access to price information and the registration, enforcement, and monitoring of rights.

Developing countries characteristically suffer from weak institutional frameworks, such as slow judicial systems, unclearly defined land rights, strong concentration of land ownership, and unenforced laws and environmental regulations. Before the introduction of water markets, the necessary legal and institutional requirements in each site should be compared with what already exists.

Socioeconomic asymmetries need to be considered because of the different types of water users, who will vary in their educational background, culture, and economic power. When water rights were allocated and made tradable in Chile, electricity companies bought up a large number of them to be held for future use, to the detriment of smaller users who at the outset did not understand the implications of their selling the rights. Given that water is the basis for agricultural production, the sale of a water right may severely undermine smallholders' livelihoods.

As in the case of land rights, poor farmers may use newly allocated water rights to pay off their debts, ending up without their entitlement in the long run. A phased approach to water trading may be the most appropriate, starting with the introduction of water rights, monitoring, and enforcement, and followed at a later stage by tradability. This would give water users-especially the more vulnerable segments of society-the time to realise the value of their new rights. Instituting a seasonal or annual leasing system could also permit right holders to experience market transactions without permanently endangering their livelihoods.

Another important issue is the impact of water reallocation on higher-valued uses and their impact on regional economies and the environment. In Mexico, small farmers with severe groundwater overdraft do not have the capital to invest in deeper wells, so they sell their water rights to large agribusinesses that do invest in wells. The small farmers win because, without the water market, they could neither invest and nor receive any compensation. Furthermore, the company creates local and regional jobs. On the other hand, by obtaining more water rights, the company also strongly contributes to the declining groundwater level, thus hurting more small farmers and the environment.

Control over and use of water has been regulated in diverse ways that involve bundles of rights. These bundles assign legitimate authority and the obligations to control water and determine the priority of water use. They lay down who has the right to appropriate water, whether or not water can be transferred, and the relation between water rights and land rights. Such bundles of rights range from the most exclusive forms of individual ownership, to communal rights at the local-community level, to public regulation at the national or state level, to agreements at the international level, or a combination of these.

Usually, any set of regulations regarding a particular water resource involves both private and public rights and obligations. The specific character of water resources requires forms of organisation that often transcend or cut across the ordinary administrative boundaries of local community, district, and state. A command area of an irrigation system may lie in more than one village; a river basin may run through several countries.

In many countries, the state claims sovereignty and ownership over all its water resources. However, especially in developing countries, state law is not the only source of regulation. Local regulations (or "customary" law) and religious regulations often assign rights and obligations that differ from and sometimes contradict state law. By definition, customary legal orders are based on different notions of who may control, regulate, and have access to water. The state does not always recognise non-state legal orders and their ideas about water rights as valid. Nevertheless, they continue to exist and exert their influence on water-resource management practices.

In assessing the nature and function of customary legal rights, policymakers often make three mistakes. First, many assume that all norms and regulations that do not emanate from state institutions are customary and long established. In fact, many rules and regulations emerge locally as new responses to outside intervention by the state or other institutions. Second, policymakers assume that everybody acts according to these rules because they are deeply ingrained in local society. However, human behavior never fully corresponds to norms and regulations. Every property regime allows for variations in behavior, and each local community adapts its legal system and the structure of rights to changing social and economic conditions. Third, policymakers assume that all customary property is communal property. In fact, no property regime is fully communal; it is always a combination of communal and individual elements, and of public and private elements.

The focus of interest shifted from economic development to sustainable development in the late 1960s in the debate about appropriate management of "the commons." Communal property rights were blamed at that time for unsustainable resource use. Communities, community-based rights, and communities' customary law are now being promoted as inherently conducive to sustainable resource use. The arguments are particularly powerful in relation to natural resources, such as water, forests, fisheries, and natural reserves.

Customary law is also often considered to be more equitable than other legal regimes. However, there is little consensus on how equity is defined. Customary legal systems may show considerable inequalities based on class, gender, age, and caste. One important reason is the unequal distribution of land. Rights to water are usually closely connected to land rights. Springs are often owned by whoever owns the land on which the spring originates. People living along a river are entitled to draw water from the river.

In cases of irrigation, the rights to irrigation water are usually distributed among those having land in the command area; or they belong to those who participate in building the system, again usually people that own the land to be irrigated. Where land is unequally distributed, water is also usually unequally distributed. Poor people tend to have land at a disadvantaged position within irrigation systems or even outside the command area of irrigation systems.

Water is also distributed unequally by gender. For their water rights, women often depend on men: fathers or brothers while unmarried, husbands after marriage. Thus, women are excluded from the right to make decisions in public arenas. Interests that tend to be more specific to women, such as accessibility to household and drinking water, are often disregarded. This is not so much a result of a conscious exclusion from water rights as such, but of more general gender differences that do not allow women to participate in the public domain, or that do not allow women to inherit or acquire property independently. State water regulations may also show inequalities. Present state policies to create water markets, meant to increase efficiency, will certainly increase inequality unless very serious measures are taken to ensure that the poor have access to this market.

Policy Implementations

There may be very good reasons for wanting to change customary water rights, for example, to bring about more efficiency, sustainability, and gender equality, or to alleviate poverty. In addition, major changes in the scope of water redistribution cannot be achieved at the relatively small political and geographical scale on which customary legal systems operate. Change should build upon a realistic and careful assessment of existing rights to water and other natural resources in all their complexity. Because rights to water are intimately linked to a wider set of social relationships, successful change requires a full analysis of existing inequalities within a society.

Any policy that simply ignores existing rights-whether defined in state law, in local legal systems, or in religious law-is bound to fail and to create more, rather than less, insecure rights. This is potentially harmful for the people that policymakers intend to protect. Policies based on false assumptions about the social working of law inevitably lead to disappointment. Replacing customary water rights with a new property regime designed from scratch is bound to meet with strong opposition from those threatened with loss of their existing rights. Changing water rights alone may only marginally redress inequity in access to water. Making changes in inheritance regulations or land redistribution may also be required.

Devising a new set of categories of rights to water will not suffice unless attention is also given to the actual distribution of water through these rights. This involves deciding on the distribution of available water resources among different uses and setting priorities. It also involves distribution decisions between different segments of the population within sectoral uses, and the legally valid means through which such redistribution is to be effected. The legal complexity of this issue and its potential for social and political tension also demand a policy style that is open to hearing and negotiating with those holding rights to water under multiple legal orders. Perfect legal regulation is likely to be an illusion, and compromises will have to be made.

Local and customary rights vary considerably in content and function from region to region. Policymakers are well advised not to cover them under uniform legislation that does not take these variations into account. New legislation should lay down a general framework and leave room for the elaboration of local variations in specific cultural, legal, and hydrological conditions.

References

Critchley, W. and Siegert, C., "Water Harvesting Manual", *FAO Paper* AGL/MISC/17/91, FAO, Rome 3, 1991.

Oweis, T., A. Hachum, and J. Kijne, "Water harvesting and supplementary irrigation for improved water use efficiency in dry areas", *SWIM Paper 7*, Colombo, Sri Lanka: International Water Management Institute, 1999.

Siegert, C., "Socio-economics of water harvesting", *Water Harvesting for Improved Agricultural Production.* Expert Consultation, Cairo, Egypt 21-25 Nov. 1993.

CHAPTER 13

Water Harvesting: The Implications of Globalisation

Global consumption of water is doubling every 20 years, more than twice the rate of human population growth. At present more than one billion people on earth lack access to fresh drinking water. By the year 2025 the demand for freshwater is expected to rise to 56% above what currently available water can deliver, if current trends persist. To solve the growing water crisis, the solution that is proposed and pushed by world bodies such as the WTO and International Monetary Fund IMF through international agreements such as GATS is privatisation of water, which in effect leads to treatment of water as a commodity.

A growing movement of people believe that the imperatives of economic globalisation-unlimited growth, a seamless global consumer market, corporate rule, deregulation, privatisation, and free trade-are the driving forces behind the destruction of our water systems. These must be challenged and rejected if the world's water is to be saved.

Impact of Economic Globalisation

Developing countries have restructured their economic systems to pay their debt and export their way to prosperity, destroying both natural ecosystems and environmental regulations. Economic globalisation has also resulted in the exponential increase in the use of fossil fuels, dams and diversions, massive transportation systems needed to carry out global trade, and roads carved out of wilderness. In the new economy, everything is for sale, even those areas of life once considered sacred, like seeds and genes, culture and heritage, food, air, and water.

Throughout Latin America and Asia, massive industrialisation in rural communities is affecting the balance between humans and nature. Water use is being diverted from agriculture to industry. Huge corporate factories are moving up the rivers of the Third World, sucking them dry as they go. Agribusinesses growing crops for export are claiming more of the water once used by family and peasant farmers for food self-sufficiency. The global expansion in mining and manufacturing is increasing the threat of pollution of underground water supplies and contaminating the aquifers that provide more than 50 percent of domestic supplies in most Asian countries.

To feed the voracious global consumer market, China has transformed its entire economy, massively diverting water use from communities and local farming to its burgeoning industrial sector. As the big industrial wells consume more water, millions of Chinese farmers have found their local wells pumped dry. Eighty percent of China's major rivers are now so degraded, they no longer support fish. Economic globalisation and the policies that drive it are proving to be totally unsustainable.

Role of Transnational Corporations

This leads to a second area of potential disagreement, the role of transnational corporations in determining the future of water. Just as governments are backing away from their regulatory responsibilities, giant transnational water, food, energy, and shipping corporations are acquiring control of water through the ownership of dams and waterways. These corporations are gaining control over the burgeoning bottled water industry, the development of new technologies such as water desalination and purification, the privatisation of municipal and regional water services, including sewage and water delivery, the construction of water infrastructure, and water exportation.

The goal is to render water a private commodity, sold and traded on the open market, and guaranteed for use by private capital through global trade and investment agreements. These companies do not view water as a social resource necessary for all life, but an economic resource to be managed by market forces-like any other commodity. At conferences like the Stockholm Water Symposium, transnational water companies assert that they are in this business for almost altruistic reasons. It is to their benefit to blur the lines between government and the private sector, and they certainly are doing a very good job of that. A closer and well-documented examination of their practices tells a very different story: higher customer rates, dramatic corporate profits, corruption and bribery, lower water quality standards, and overuse of the resource for profit.

While the companies argue that the privatisation of water services is socially beneficial, the consequence of corporate control is that social and environmental concerns come second to the economic imperative of maximum profits for the shareholders.

Global Water Corporation of Canada has contracted to ship 58 billion litres per year of Alaskan glacier water by tanker to be bottled in a free trade zone in China. They openly boast that the venture "will substantially undercut all other imported products" because of China's cheap labour. Global Water entices investors to "harvest the accelerating opportunity as traditional sources of water around the world become progressively degraded and depleted" and declares that "water has moved from being an endless commodity that may be taken for granted to a rationed necessity that may be taken by force."

Privatising Water

Privatising water is the best way to deliver it safely to a thirsty world. This is yet another area of potential disagreement. It is true that governments have done an abysmal job of protecting water within their boundaries. The privatisation of water ensures that decisions regarding the allocation of water centre almost exclusively on commercial considerations. Corporate shareholders are seeking maximum profit, not sustainability or equal access. Privatisation means that the management of water resources is based on the principles of scarcity and profit maximisation rather than long-term sustainability. Corporations are dependent on increased consumption to generate profits and are much more likely to invest in desalination, diversion, or export of water rather than conservation.

The global trend to commodify what has been a public service reduces the involvement of citizens in water management decisions. For example, private water projects brokered by the World Bank have minimal disclosure requirements. A water corporation executive at the recent World Water Forum in The Hague, said publicly that as long as water was coming out of the tap, the public had no right to any information as to how it got there. The concentration of power in the hands of a single corporation and the inability of governments to reclaim management of water services allow corporations to impose their interests on government, reducing the democratic power of citizens.

As the South African Municipal Workers' Union explained, "Water privatisation is a crucial issue for public debate. Human lives depend on the equitable distribution of water resources; the public should be given a voice in deciding whether an overseas-based transnational corporation whose primary interest is profit maximisation should control those critical resources. Water is a life-giving scarce resource that must remain in the hands of the community through public sector delivery. Water must not be provided for profit, but to meet needs."

Advocates of privatisation argue that they are seeking private-public partnerships, and give assurances that governments will still be able to establish regulations. Since the provision of water services itself does not provide sufficient return, water corporations are actively pursuing exclusive control over water service provision through acquisitions of infrastructure and water licenses. They are closing the loop around public involvement and creating huge monopolies against which local suppliers cannot compete.

In their support for large-scale project financing, the World Bank and others give preference to large multi-utility infrastructure projects that favour the biggest corporations, leading to monopolies. To add insult to injury, the World Bank underwrites these giant corporations with public money, and often incurs the risk, while the companies reap the profit. And often, governments, who supposedly represent their people, have to assure a return to the shareholder. Most disturbing, the close alliance between governments, the World Bank, the United Nations, and the water companies gives these corporations undue influence over government policies that favour their interests, like deregulation and free trade, and favoured access to upcoming water contracts. The stated goal of the World Bank water loan to Budapest was to "ease political resistance to private sector involvement."

Pricing

Pricing water will cause to understand its real value and force to start conserving it from economic necessity. Water pricing exacerbates the existing global inequality of access to water. The countries that are now suffering severe water shortages are home to the poorest people on earth. To charge them for already scarce supplies is to guarantee growing water disparities. The issue of water pricing will also exacerbate the North/South divide. There is a sub-text inherent in much of the hand-wringing over the world's water shortage. The privatisation of this scarce resource will lead to a two-tiered world-those who can afford water and those who cannot. It will force millions to choose between necessities such as water and health care. In England, high water rates force people to choose whether or not to wash their food, flush their toilets, or even bathe.

Water pricing, combined with privatisation, will seal water's fate as a commodity under the terms of international trade agreements supported by the World Trade Organisation (WTO) and the North American Free Trade Agreement (NAFTA). Both the WTO and NAFTA consider water to be a tradeable good, subject to the same rules as any other good. Only if water is maintained as a public service, delivered and protected by governments, can water be exempted from the onerous enforcement measurements of these trade deals. Claiming environmental exemptions for water will not suffice. Every single time the WTO has been used to challenge a domestic environmental rule, the corporations have won and the environmental protection has been ruled 'trade illegal.'

The trade agreements are very clear: if water is privatised and put on the open market for sale, it will go to those who can afford it, not to those who need it. By the terms of trade rules, once the tap has been turned on, it cannot be turned off. The World Bank says that it will subsidise water for the poor. Anyone familiar with the problems of welfare, particularly in the Third World, knows that such charity is punitive at best, and more often, non-existent. Water as a fundamental human right is guaranteed in the UN Universal Declaration on Human Rights. Water welfare is not what the architects of that great declaration had in mind.

Another argument against pricing is that, as it is now envisaged, it won't have much of an impact. It is generally accepted that water consumption in urban centres breaks down at 70 percent industrial, 20 percent institutional and from six to ten percent domestic. Yet most of the discussions about water pricing are around individual water use. Large corporate users notoriously evade the cost of their water altogether.

For example, in California's Silicon Valley, the high tech sector uses huge amounts of water. This sector is presently engaged in mechanisms to capture traditional water rights: *water pricing*, whereby industry pressures governments for subsidies and circumvents city utility equipment to directly pump water, thus paying much less than residential water users pay for water; *water mining*, whereby companies gain rights to deplete the aquifers while driving up the access costs to smaller users such as family farmers; *water ranching*, whereby industry buys up water rights of ranches and farmers; and *water dumping*, whereby industry contaminates the local water sources and then passes the costs on to the community.

Water Privatisation in India

Water privatisation is increased cost for water will promote conservation. This commodification of water has already happened in several developed countries and is being pushed in many developing countries through structural adjustment policies. The control of water by private companies takes away this resource from the public and puts it in private control. Privatisation of water is a violation of basic rights of citizens and that privatisation creates water monopolies and brings about more disparity in access to water.

India can be broadly divided into 15 ecological regions. The vast ecological diversity of this country is reflected in the diversity in available water resources. With an average annual rainfall of 1,170 mm, India is one of the wettest countries in the world. However, there are large variations in the seasonal and geographical distribution of rainfall over the country. At one extreme are areas like Cherrapunji, in the northeast, which is drenched each year with 11,000 mm of rainfall, and at the other extreme are places like Jaisalmer, in the west, which receives barely 200 mm of annual rainfall. Though the average rainfall is adequate, nearly three-quarters of the rain pours down in less than 120 days, from June to September.

India's groundwater resources are almost ten times its annual rainfall. According to the Central Groundwater Board of the Government of India, the country has an annual exploitable groundwater potential of 26.5 million hectare-meters. Nearly 85% of currently exploited groundwater is used only for irrigation. Groundwater accounts for as much as 70-80% of the value of farm produce attributable to irrigation. Besides, groundwater is now the source of four-fifths of the domestic water supply in rural areas, and around half that of urban and industrial areas.

According to the International Irrigation Management Institute (IIMI), the water table almost everywhere in India is falling at between one to three meters every year. Furthermore, the IIMI estimates that India is using its underground water resources atleast twice as fast they are being replenished. Already, excessive ground water mining has caused land subsidence in several regions of Central Uttar Pradesh.

There are 14 major, 44 medium and 55 minor river basins in the country. The major river basins constitute about 83-84% of the total drainage area. This, along with the medium river basins, accounts for 91% of the country's total drainage. India has the largest irrigation infrastructure in the world, but the irrigation efficiencies are low, at around 35%.

If per capita water availability is any indication, 'water stress' is just beginning to show in India. This index is based on the minimum per capita level of water required to maintain an adequate quality of life in a moderately developed arid zone country. A region where renewable fresh water availability is below 1700 cubic meters/capita/annum is a 'water stress' region, and one where availability falls below 1000 cubic meters/capita/annum experiences chronic 'water scarcity'. The annual per capita availability of renewable freshwater in the country has fallen from around 5,277 cubic meters in 1955 to 2,464 cubic meters in 1990. Given the projected increase in population by the year 2025, the per capita availability is likely to drop to below 1,000 cubic meters i.e., to levels of water scarcity.

Demand and Resource Management

Water availability on the Indian subcontinent is strongly influenced by a number of climatic and geographic factors. Together these combine to provide India with enough freshwater to meet the various demands arising from the agricultural, industrial and domestic sectors. However, the actual distribution of water resources over space and time limits access to certain geographic regions and during a few months of the year. Government policies and economic incentives have also influenced the water distribution and consumption across India.

Ancient Indian history is full of such references to techniques for harvesting all forms of water - rainwater, groundwater, surface water such as streams and river and even floods. These water harvesting structures were based on traditional models which had been refined over centuries to adapt to the prevailing climatic and hydrologic conditions in their area of use. The potential of such water harvesting systems to supply adequate freshwater to all areas and sectors is high. However, since colonial times—and especially after independence in

1947—these systems have been increasingly abandoned and neglected in favour of large dam and canal irrigation projects. So far, these 'temples of modern India' have been successful in providing water to selected portions of rural and urban India; yet high economic, social and environmental costs have reduced their overall benefit.

The highly variable nature of rainfall makes groundwater the most popular alternative for irrigation and domestic water use across India and accounts for over 400 km^3 of the annual utilisable resource in the country. This dependence on groundwater resources is particularly critical where dry season surface water levels are low or where wet season flows are too disruptive to be easily tapped. In addition to being accessible, groundwater quality is generally excellent in most areas and presents a relatively safe source of drinking water for Indians in rural and urban centres.

Agriculture remains central to the Indian economy and it therefore receives a greater share of the annual water allocation. According to the World Resources Institute, 92% of India's utilisable water is devoted to this sector, mostly in the form of irrigation. Groundwater alone accounts for 39% of the water used in agriculture and surface water use often comes at the expense of other sectors such as the industrial and domestic supply. Demand from the domestic sector has remained low and accounts for only 5% of the annual freshwater withdrawals in India.

The demand from domestic sector over the next twenty years will increase from 25 billion m^3 to 52 billion m^3. However, this increase in the demand from the domestic sector will not be as much as that from other sectors over the next several years. Currently, only 85% of the urban and 79% of the rural population has access to safe drinking water and fewer still have access to adequate sanitation facilities.

Recognising that the growing demand for water in agriculture and industries sets a pattern of water scarcity even in areas where there is sufficient water for domestic purpose, the national water policy has rightly prioritised drinking water over other uses. However, in giving subsidies to the industrial and agriculture sectors where the water consumption is highest and allowing these sectors to use more water at negligible prices, the government has effectively contradicted its own water policy. This has resulted in mining of ground water leading to a rapidly falling water table. For example, the bottling companies of Pepsi and Coca-Cola in different parts of India pay very little towards water mining and have practiced unsustainable water mining in these areas to the detriment of villagers and small farmers in the vicinity.

Urban areas are serviced by a municipal water distribution system. Usually, the municipal water supply originates from local reservoirs or canals, but in some cases water may be imported through inter-basin transfer. Although the major cities in India enjoy access to central water supply systems, these schemes often do not adequately cover the entire urban population and are notoriously inefficient and unreliable. In rural areas, access to water is even more precarious. Over 80% of the rural domestic water comes from groundwater sources since it is more reliable in terms of water quantity and quality.

In the past several decades, industrial production has increased in India owing to an increasingly open economy and greater emphasis on industrial development and international trade. Water consumption for this sector has consequently risen and will continue growing at a rate of 4.2% per year. According to the World Bank, demand of water for industrial, energy production and other uses will rise from 67 billion m^3 to 228 billion m3 by 2025. The United Nations has warned that by 2025 two-thirds of the world will face severe water shortages if the current pattern of water consumption continues. Water-privatisation has been sought as one of the solutions to address the looming crisis.

Privatisation and Management

Water privatisation involves transferring of water control and/or water management services to private companies. The water management service may include collection, purification, distribution of water, and waste water treatment in a community. Traditionally this service has been provided by the local governmental infrastructure such as the municipality or local city council. The pro privatisation lobby including water corporations, the World Bank and IMF has aggressively campaigned for water privatisation on the grounds that, while water subsidies promote wasteful practices, commodification of water should allow market forces (supply and demand) to set the water tariff, which in turn will reduce water consumption and promote water conservation.

Furthermore, it is argued that opening this sector to private providers, will bring in badly needed capital for upgrading and development of infrastructure. There are several models of water privatisation that are currently in vogue in different parts of the world. Water privatisation has been recommended by the Indian government's national water policyto address the issue of water scarcity. In its article 13 titled, "Private sector participation" the policy says that "private sector participation should be encouraged in planning, development and management of water resources projects for diverse uses, wherever feasible". This has placed water privatisation at the forefront of developmental policies implemented by several state governments.

The national water policy also encourages interlinking of rivers to improve water availability in water scarce areas. The proposed river linking scheme has water privatisation at its heart of funding, which will further isolate the water source and responsible water management from local communities. Many state governments, neighbouring nations sharing river waters with India and experts have questioned the merits of such a scheme on numerous grounds including lack of feasibility and impact studies on this project, ecological disasters from river diversion schemes around the world, as well as adverse environmental impact due to submergence, soil salinity and water logging.

The privatisation that has been implemented in USA, UK and Canada exemplify the issues with privatisation. Several cities have opted not to privatise their water operations. In UK, privatisation came at with huge government subsidies. The government converted the regional water authority bodies into private profit making entities, wrote off $8 billion in debt of the water companies before privatisation and gave a further $2.6 billion to help the newly formed companies meet new EU environmental requirements. To complete the deal, the new companies were guaranteed private monopolies for 25 years and were given a special exemption from paying taxes on their profits. Following privatisation, there was an unprecedented water price hike of over 50% in the first 4 years; the corresponding rate of household disconnections due to non-payment also rose three-fold.

In Walkerton of Ontario, Canada, water privatisation has not been regulated sufficiently and E.Coli outbreak from contaminated water has caused several deaths in the region. Several bottling operations and transfer of control of water assets such as lakes to water corporations have been stiffly opposed by local population. In 1994 the city of Hamilton, Canada privatised its water and sewerage utility works expecting it to be used as a "model" for other cities to open up their public utilities to privatisation. However, the fiasco that ensued privatisation in Hamilton made it a model for what to avoid. Over a period of six years after privatisation, residents saw more sewage spills, environmental fines left unpaid for years, and rising tariffs for water services.

Water privatisation has invariably led to price hikes in almost all the regions in the world where water has been privatised. This is because there are considerable costs involved in upgrading water harnessing, purification and distribution systems. In developing countries such as India, the water price hike is also an indirect consequence of the conditions imposed on the governments by the World Bank and IMF in return for structural adjustment loans. The privatisation of public services such as water and electricity is often a sine qua non for such loans. Furthermore, full cost recovery is demanded by the World Bank and IMF as a prerequisite to privatisation. For instance, during the severe flooding in Orissa in 2001, the World Bank demanded an increase in the water tariffs as a cost recovery measure on the use of water.

Rates for water irrigation have since doubled or even tripled. Increased consumer fees for water can make safe water unaffordable for the impoverished and vulnerable populations. Families are often forced to make trade-offs between water, food, schooling and health care. These cost recovery conditions mean that user fees paid by water consumers must cover all water system costs, which usually include the costs of operation, maintenance and capital expenditure, and sometimes the cost of servicing past utility company debt.

The World Bank justifies cost recovery requirements by contending that, with higher payments from consumers, private companies will have an incentive, as well as the revenues, to extend pipes to those relying on water trucks or unclean sources. However, there is little evidence of the multinational water companies' commitment to expanding service, especially to poor communities where the ability to pay increased fees is limited. This is because the poor communities offer little or no margins to the water corporations. Instead, the multinationals, which have only recently started their major moves into developing countries, have quickly racked up very poor social and environmental records.

Many potential risks emerge once a resource as fundamental to life as water is privatised. One of the foremost reasons to oppose water privatisation is the threat of unsustainable water mining by the water corporations in an effort to maximise profits. These corporations, which are answerable only to their shareholders, have a declared agenda to make profit. Once water becomes a marketable commodity and a corporation is given sole rights to a body of water, then it is within the corporation's rights to mine as much water as it deems fit. Furthermore in an effort to maximise profits, if the corporation mines an environmentally unsustainable amount of water and deplete the water body at a rate faster than it is replenished, then the government officials and the affected population can do very little to legally prevent the corporation from doing so.

The fact that this is a real and tangible threat is apparent from the increasing number of community complaints against indiscriminate mining of groundwater by Coca-Cola in the Khammam district of Andhra Pradesh, Athur village near Chennai and Plachimada in Kerala. Residents from villages in the Palghat district in Kerala surrounding Coke's Greenfield softdrink bottling factory in Plachimada say that Coke's indiscriminate water mining has dried up many wells and contaminated the rest. Coca Cola's bottling plant was set up in 1999 in the middle of fertile agricultural land, with proximity to a number of reservoirs and irrigation canals. Coke's mining of more than 1 million liters of ground water per day has parched the lands of some 2000 people within 1.2 miles of the factory.

A water corporation can use one of the many free trade treaties to take legal action against the government for withdrawing from an agreement to privatise water systems. Although this has not yet happened in India, there are numerous instances of lawsuits filed by water corporations against local governments that backed out of a contract in countries with a longer history of water privatisation. Unlike privatisation of other sectors such as airlines or telecommunications, privatisation of water services (and other essential services) often does not leave the consumers with a choice of provider.

Physical reliance on a single water pipe network (and often a single water source) leaves little room for competition, which lends monopolistic attributes to privatisation in this sector. This underscores the need for a highly competent, well-funded and politically autonomous regulatory body before privatisation is undertaken. Yet in the poorest countries where privatisation is promised to bring the greatest benefit, these institutional preconditions are almost always missing. Furthermore, the water corporations demand exclusive service provider rights for periods of 5-20 years claiming long recovery periods for costs involved in upgrading water infrastructure.

Corporations in search of profits can compromise on water quality in order to reduce costs. This is especially true in a country such as India, where the water quality regulatory boards do not have the teeth to enforce their standards. There have been numerous instances of outbreak of epidemics due to poor quality of water.

In India, the bottled water industries and cola industry have been shown to have high pesticide levels in their products. These are classic examples of how private companies cannot be expected to provide quality with out regulation. However privatisation pushed by IMF and the World Bank is based on free market economy and regulation of any such measure is opposed by these world bodies. Hence while privatisation of essential services is pushed, regulation of these services is being removed through budget cuts.

Indian government agencies are notorious for their lack of accountability and transparency in awarding of service contracts to private corporations. Government management of water resources is partly responsible for the water crisis we face today, privatisation will at best compound the problem. There is a clear need for alternate models which offer equitable, sustainable and efficient management of water resources. Such models exist all over the world and have been quite successful in their implementation and public acceptance.

In contrast to privatisation, there are several community based small-scale water management and distribution models, which offer cost benefits and long term sustainability over the privatisation models. Participatory and co-operative models deliver impressive results by ensuring that water utilities are accountable and responsive to the needs of the local communities. Furthermore, while these models address the same global issue of water scarcity, they tackle it with local solutions tailored to their specific environments.

The harvesting of rainwater involves the collection of water from surfaces on which rain falls, and subsequently storing this water for later use. Normally water is collected from the roofs of buildings and stored in rainwater tanks or allowed to recharge ground water and thus providing a perennial source of water. In India, with its monsoons, rain water harvesting experience is millennial, born of local wisdom and is widely used today. It is scientific and the focus of innovation in the present, participatory and the basis of people's movements, and the best way to a non-scarce future.

Rain water harvesting is increasingly being modeled and implemented successfully in urban centres where the likelihood of privatisation is the maximum. Traditionally, Chennai city in South India has been dependent on groundwater to a large extent for its water needs. However, large-scale construction activity has led to rapid depletion of the groundwater. Chennai gets an average annual rainfall of 1,290 mm per year.

Another case is of New Delhi, where as the city's water table plummets to dangerously low levels, the government is making desperate efforts to stave off the water crisis. Having tried it in the chief minister's residence, the government has now made rainwater harvesting mandatory for every new house or hotel being built in the Capital. The Delhi Development Authority and the Municipal Corporation of Delhi have made amendments in the existing building bylaws, making it compulsory for every house or hotel, which is 200 yards and more in area, to undertake rainwater harvesting.

In Gujarat, Utthan, a non-government organisation, started gender empowerment development work in the region of Bhal, it quickly realised that access to water is a key issue that needs to be tackled for women empowerment. This region in Gujarat suffers from a hostile geo-climatic environment, highly saline shallow ground water and erratic monsoon rains. Through a highly effective community mobilisation and organisational work, Utthan and Mahiti (a sister organisation) were able to initiate women's movement in Bhal focused around the issue of access to safe and regular supply of drinking water. This movement graduated to create such a powerful pressure on the local and state level bureaucracy that even the hardened policy makers had to sit back and take a serious notice of it. These women in Bhal pressurised the Gujarat Water Supply and Sewerage Board (GWSSB) to approve a project that sought to promote highly decentralised rain water harvesting structures such as plastic lined ponds and roof top water collection tanks in the villages of Bhal. The overall contribution of Utthan and Mahiti in Bhal, therefore, can be viewed in terms of their having initiated a social movement in this area which focused on community managed drinking water resource development within a gender sensitive framework.

Kothapally village in the Ranga Reddy district of Andhra Pradesh is blessed with rich black soil but little rain. The village's watershed management programme, which is run by a consortium of partners under the leadership of the international crop research institute (ICRISAT), has become a model for community involvement in watershed development. The consortium includes the Government of Andhra Pradesh, the Drought Prone Area Programme, the Central Research Institute for Dryland Agriculture, the National Remote Sensing Agency, and the Farmer's Association. Under this project, several techniques for soil and water conservation and crop improvement were implemented to the benefit of the small farmers. These farmers saw the value of water harvesting and joined the consortium of project partners in constructing 13 check dams and 51 mini percolation tanks, which has increased water levels, expanded green cover and enhanced the productivity of crops.

Tarun Bharat Sangh (TBS), a non-government organisation, has been working wonders in drought-prone Rajasthan, where agriculture output has doubled and forest cover increased substantially. TBS started mobilising villagers to reconstruct a damaged Johad (dam) at Gopalpura. The 426 meter Johad was reconstructed in the next monsoon season. It took the villagers almost a year to complete renovating the Johad. When the Johad was finally renovated, more than 600 "Bighasi" of land became suitable for agriculture. TBS now helps villages to build water harvesting systems. So far, 4,500 Johads have been built in 850 villages covering 6,500 sq kilometers in the last 15 years.

According to Rajendra Singh (founder of TBS), they only propagate the idea of water harvesting through Pani Yatras, where they select a group of villages to spread the word about this system. They make the villagers aware about water harvesting. When a village approaches TBS, they tell them what the whole system is about. Then it is up to the villagers to draw up plans and implement the project. Once a village decides to implement the project, a meeting of the gram sabha is held to discuss the details of the project. Every villager contributes when funds are collected in cash or kind for the project. If there is a dearth of money, the TBS contributes up to 67 per cent of the required funds. The villagers have to pay a minimum of 33 per cent for the project.

References

Alex Kirby, "World Water Shortage Threatens Harvests," BBC World News, February 7, 1999

Ann Ninan, "Private Water, Public Misery," *CorpWatch India*, April 16 2003

David Hall, "Privatization, Multinationals and Corruption", *Public Services International Research Unit*, Report No. 9909-U-U-Corrup.doc, July 1999.

Falkenmark, Malin and Carl Widstrand, "Population and Water Resources: A Delicate Balance," *Population Bulletin*. Population Reference Bureau, 1992

Ghotge, Sanjeev, "India's Water Policy – A Perspective Paper," National Centre for Advocacy Studies, Pune, October 2002.

Keya Acharya, "India's River Linking Scheme Worries the Region's Journalists," Planets-Voice.org, July 14, 2003.

Naunidhi Kaur, "Privatising Water," *Frontline*, vol. 20, issue 18, Aug 30- Sept 12, 2003

Nityanand Jayaraman, "No water? Drink Coke", *CorpWatch India*, May 28, 2002,

Sonu Jain, "A River gone private is drying up" *The Sunday (Indian) Express*, April 13, 2003

Sudhirendar Sharma, "Water Resources: Background and Perspective," InfoChange News and Features, April 2003

Vandana Shiva, "Water Privatization in India," *Article in Public Citizen website*, July 31, 2002

CHAPTER 14

Integrated Water Harvesting Management

Integrated River Basin Management (RBM) systems bring fragmented water uses and users together. RBM creates a framework that deals with an entire basin or sub-basin. A river basin can be defined by the watershed limits of a system of waters, both ground and surface, flowing to a common terminus. RBM integrates this system of waters within its broader natural environment and its social, economic, and political contexts. Basin units cut across administrative divisions used to manage water. This is their strength and their challenge.

Water in a river basin is both renewable and reusable. The hydrologic cycle renews the resource annually; water within a basin can be reused many times on its way from headwaters to the sea. Some are in-stream uses, such as hydropower generation, navigation, recreation, or ecological sustenance. Other uses, such as municipal water supply, industrial cooling, or irrigation, extract water but return much of it downstream. As a result, a blend of water that has been used and reused comprises the lower reaches of most major river systems. This multiplier effect allows withdrawals to exceed by many times the natural flow in a river and provides a number of management options for optimising the utility of water in a basin, but creates significant measurement and quality management problems.

Because of the large variability in river flows across seasons and years, flood protection works and artificial storage are important elements of river basin systems. The degree of control afforded by reservoirs varies enormously. The reservoirs on the Colorado River in the U.S. can store four times the annual flow of the river, whereas the Mekong River in Southeast Asia is largely unregulated. There are many more management options and the value of water in the Colorado River basin is higher, since regulated flows generally have much greater utility for human uses than unregulated ones.

Integrated basin management can also encompass a number of other divisions, including the following:

i) Administrative jurisdictions;
ii) Ground and surface water;
iii) Upstream and downstream reaches;
iv) Environmental and human uses;
v) Supply and demand management;
vi) Water quantity and quality;
vii) Land and water use; and
viii) Transboundary uses.

This value comes at a cost to natural ecosystems, which have evolved from and adapted to natural cycles of flood and recession. Manipulating reservoir releases to mimic key portions of these natural cycles has increased, but much remains to be learned in this area. As river basins approach full utilisation, systems of specifying, quantifying, and assigning rights to the water become more formalised. Those affected try to protect, defend, and perhaps augment their rights; interact with other right holders, claimants, and aspirants; and sometimes conflict with them.

RBM must therefore establish the rules of the game and mechanisms to govern these interactions, so that they take place in a productive and efficient way. All interests and stakeholders in the water affairs of the basin must have suitable protection and adequate representation, including the natural environment and less powerful water users. The most important decisions in RBM are made on the basis of contemporary social values, not on technical grounds. Effective RBM requires both good process and good science.

Organisational Models for Implementing RBM

There are two archetypal organisational models for implementing RBM. The first is the authority model, in which a single unified organisation is empowered to make decisions. The second is the coordinative model, in which existing administrative units work together to cover an entire river basin or subbasin. While new structures may be created, the bulk of routine work is done by existing organisations.

A strength of the authority model is that its operational span of control coincides with the boundaries of the basin. This internalises upstream/downstream and other conflicts, making them easier to deal with. And it concentrates the decisionmaking authority needed to resolve disagreements. But there are disadvantages too. Since the authority will very likely deal only with water, water will be isolated from relevant policy sectors such as agriculture, the environment, and the economy. Authority is centralised rather than devolving to the lowest practicable level. And for international rivers, the authority model requires establishing a supranational authority, which is extremely difficult to do. Finally, governance of an RBM authority may not include broad-based stakeholder representation and accountability.

The coordinative model addresses some of these weaknesses. Because coordination involves voluntary agreement among participating jurisdictions, it provides a strong political base for action. Linkages between water and other types of policy remain, since states, nations, or other jurisdictions are jointly responsible for a range of policy sectors. Such a set-up also provides a natural base for decentralisation of responsibilities. On the other hand, decisionmaking can be cumbersome, costs of coordination may be high, and political changes in participating jurisdictions can upset agreements. These two models represent polar extremes.

In Australia's Murray-Darling Basin, a cooperative Ministerial Council comprising representatives of the four involved states and the federal government, sets policy while an authority-like commission supports and executes the council's decisions. In France, a river commission made up of local and national government representatives and users sets water policy, which an associated water agency implements. Publicly held companies manage the distribution infrastructure and make bulk water deliveries to user associations. In the United States, there is generally no formal apex council in a river basin, and policymaking authority is distributed among federal and state agencies and departments.

Legislation and legally binding negotiated agreements are important instruments for establishing policy and practices, and the court system resolves disagreements and disputes. In the state of California, a water plan that is updated every five years provides a rolling framework for managing the state's water resources.

RBM in developing countries and in developed countries may differ because in developing countries:

i) Dense populations live in upper watershed catchments and require access to water along with downstream urban populations;

ii) Water use is widely dispersed rather than concentrated and easily controlled;

iii) Administrative capacity to monitor and enforce regulations and standards is limited;

iv) Governance mechanisms to assign rights and regulate and enforce agreements and contracts are nonexistent or weak;

v) Technical capacity to measure and monitor basin hydrology is limited;

vi) Civil society does not possess the internal structure of groups and associations needed to represent various stakeholder interests;

vii) Pressures for transparency in public decisionmaking and regulation, including independent, investigative media, are not strong.

These differences preclude the wholesale importation of developed-country models. RBM strategies must be adapted to the particular context. Building institutional capacity in critical areas may have to proceed simultaneously with the development of RBM organisations. In such cases, it may be necessary to defer assigning particular functions to these organisations until sufficient capacity has been developed.

Achieving effective basin management requires a strong knowledge base, suitable governance mechanisms, administrative capacity, adequate stakeholder representation, transparency, and political will. This challenging set of requirements is matched by equally rewarding potential outcomes, such as more efficient use of each drop of water, enhanced reuse of water, mitigation of past environmental damage, and redirection of water to uses that society values most highly. The overarching imperative is to accomplish these ends in a way that treats all stakeholders, particularly weaker and poorer stakeholders, fairly and equitably.

References

Srinivasan, R., Arnold, J. G., "Integration of a Basin-Scale Water Quality Model With GIS", *Water Resources Bulletin* 30, no. 3: 453, 1994.

Scudder T., "Recent experiences with river basin development on the tropics", *Natural Resources Forum* 18(2): 101-114, 1994

Bibliography

Abdi, A. M., "Water harvesting systems in the Northwestern region of Somalia", *Paper presented to the World Bank Workshop on Water Harvesting in Sub-Saharan Africa*, Baringo, Kenya, 13-17. Oct 1986.

Achouri, M., "Small Scale Water Harvesting in Tunisia", *In: FAO, Water Harvesting For Improved Agricultural Production*, Expert Consultation, Cairo, Egypt 21-25 Nov. 1993, p. 87-97, FAO, Rome, 1994.

Agarwal Anil, Narain Sunita eds., "Dying Wisdom, Rise, Fall and Potential of India's Traditional Harvesting Systems, State of India's Environment", *A Citizens' Report* 4, Delhi, Centre for Science and Environment, 404 p., 1997.

Agnihotri, Y.; Samra, J.S.; Mittal, S.P., "Boosting Hill Resource Economy through Watershed Management in Hoshiarpur, Shiwaliks, India", *Indian-Journal of Soil Conservation*, 24: 3, pp 248-252, 1996.

Alex Kirby, "World Water Shortage Threatens Harvests," BBC World News, February 7, 1999

Al-Labadi, M., "Water Harvesting in Jordan- Existing and Potential Systems", *In: FAO, Water Harvesting For Improved Agricultural Production*, Expert Consultation, Cairo, Egypt 21-25 Nov. 1993.

Aminipouri, B & Ghoddousi, J., "Rainwater Catchment for Survival", *Proceedings of the 8th International Rainwater catchment Systems Conference*, Tehran: Ministry of Jihad-e- Sazandegi, 1997.

Ann Ninan, "Private Water, Public Misery," *CorpWatch India*, April 16 2003

Ariyabandu R.de.S, and Dharmalingam S., *Harvesting Rainwater: A means of Water Security to Rural Sri Lanka*, HARTI, Cololombo, Sri Lanka, 1997.

Aslam, M.; Ikram, M. Z., 'Water Harvesting and Dugwells for Sustainable Water Use', In *Progressive Farming*, 14 (5) 66-69, 1994.

Ballabh Vishwa, Choudhary Kameshwar, *Consultation Workshop on Small Water Harvesting Structures: Technology, Institution and Management*, Anand, Institute of Rural Management, 44 p., 1999.

Barrow, C., *Water Resources and Agricultural Development in the Tropics*, Longman, UK, 1987.

Ben-asher, J., "A review of water harvesting in Israel", (Draft) *Working Paper for World Bank's Sub-Saharan Water Harvesting Study*, 1988.

Bhaduri, B., Harbor, J., Engel, B., and Grove, M., "Assessing Watershed-Scale, Long-Term Hydrologic Impacts of Land-Use Change Using a GIS-NPS Model", *Environmental Management* 26, no. 6: 643, 2000.

Billy, B., "Water harvesting for dryland and floodwater farming in the Navajo Indian reservation", In: Dutt, G.R. et al (eds), *Rainfall collection for agriculture in arid and semi-arid regions: 3-8*, Commonwealth Agricultural Bureaux, UK, 1981.

Bochniarz Z., "Water management problems in economics in transition", *Natural Resources Forum* 16(1): 55-64, 19923.

Boers, T.M. and Ben-Asher, J., "Review of rainwater harvesting", *In: Agricultural Water Management* 5: 145-158, 1982.

Bond, G. *et al.*, "A pervasive millennial-scale cycle in North Atlantic Holocene and glacial climates" *Science*, 278, 1257–1266, 1997.

Bruins, H. J., Evenari, M. and Nessler, U., *Rainwater harvesting agriculture for food production in arid zones: The challenge of the African famine*, Appl. Geogr. 6:13-33, 1986.

Bruins, H.J. Evanari, M. and Nessler, U., "Rainwater harvesting agriculture for food production in arid zones: the challenge of the African famine", *In: Applied Geography* 6 (I): 13-33, 1986.

Chakraborti, D. *et al.*, "Arsenic groundwater contamination and sufferings of people in Rajnandgaon district, Madhya Pradesh, India", *Curr. Sci.*, 77, 502–504, 1999.

Chang, T., Hsu, M., Teng, W., and Huang, C., "A GIS-Assisted Distributed Watershed Model for Simulating Flooding and Inundation", *Jounal of American Water Resources Association* Vol 36, No. 5: 975, 2000.

Clarke, R., *Water The International Crisis*, Cambridge, Ma: MIT Press, 1993.

Critchley, W.R.S., "Runoff harvesting for crop production: experience in Kitui District; 1984-1986", *Paper presented to the Third National Soil and Water Conservation Workshop*, Nairobi, Kenya, 1986.

Cullis Adrian, Pacey Arnold, *Rain Water Harvesting*, London, Intermediate Technology Publisher, 1991.

Cunliffe, D. "Guidance on the Use of Rainwater Tanks, National Environmental Health Forum

David Hall, "Privatization, Multinationals and Corruption", *Public Services International Research Unit*, Report No. 9909-U-U-Corrup.doc, July 1999.

Davis S S., "Implementing gender policy in the water and sanitation sector", *Natural Resources Forum* 20(3): 189-198, 1996

Dutt, G.R., Hutchinson, C.F. and Garduno, M.A. eds., "Rainfall collection for agriculture in arid and semi-arid regions", *Proceedings of a workshop*, University of Arizona and Chapino Postgraduate College. Commonwealth Agric. Bureaux, UK, 1981.

Edwards K A, "Water, environment and development: a global agenda", *Water Resources Forum* 17(1): 59-64, 1993

Engelman R and Roy P., "Sustaining Water: Population and the future of Renewable Supplies", Washington, DC: Population and Environment Programme, Population Action International. 56 pp, 1993

Evanari, M., Shanan, L. and Tadmor, N.H., *Runoff farming in the Negev desert of Israel: 2nd Progress Report Avdat and Shivta farm project*, 1962/65. Rehovot, Israel, 1964.

Falkenmark, Malin and Carl Widstrand, "Population and Water Resources: A Delicate Balance," *Population Bulletin*. Population Reference Bureau, 1992

FAO, "Water Hervesting for Improved Agricultural Production", *Proceedings of the FAO Expert Consultation*, FAO Rome, 424 pp., 1994.

Ghotge, Sanjeev, "India's Water Policy – A Perspective Paper," National Centre for Advocacy Studies, Pune, October 2002.

Gleick, P.ed., *Water in Crisis: A guide to the World's Fresh Water Resources*, Oxford Univ.Press, 1993.

Gleick. P.H., "Basic Water Requirements of Human Activities: Meeting Basic needs", *Water International* 21, 83 - 92, 1996.

Gould, J. & Nissen-Petersen, E., *Rainwater Catchment Systems for Domestic Supply: Design, Construction and Implementation* , London: I.T. Publications, 1999.

Gupta, A. K., Anderson, D. M. and Overpeck, J. T., "Abrupt changes in the Asian southwest monsoon during the Holocene and their links to the North Atlantic Ocean", *Nature*, 421, 354–357, 2003.

Hoerling, M. and Kumar, A., "The perfect ocean for drought", *Science*, 299, 691–694, 2003.

Huibers, F. P., "Rainfed Agriculture in a Semi-Arid Tropical Climate: Aspects of Land and Water Management for Red Soils in India", *Doctoral Thesis*, Agric. Univ. Wageningen, NL, 1985.

Keya Acharya, "India's River Linking Scheme Worries the Region's Journalists," Planets- Voice.org, July 14, 2003.

König, K., *Rainwater in Cities, Ecological Concepts,* fbr, Kassler Str. 1a, D-60486, Frankfurt am Main, Germany, 1998.

Koudstal R, Rigsberman F R, Sauenije H., "Water and sustainable development", *Natural Resources Forum* 16(4): 277-290, 1992

Ladda, Brijgopal, *Planning and Financing Urban Water Supply: A Case Study of Ahmedabad,* Ahmedabad, CEPT, 1995.

Mathur O P., "The issue of Resources in the Context of growing urbanization", pp. 23-30, In R Krishnan (ed) *Growing Numbers and Dwindling Resources*, New Delhi: Tata Energy Research Institute, 171 pp., 1994

Monographs", *Water Series* 3, Public and Environmental Health Service, Dept. of Human Services, PO Box 6, Rundle Mall SA 5000, Australia, 1998.

Mou, H. et. al. Edit., "Rainwater Utilization for the World's People, Proceedings of the 7th International Rainwater Catchment Systems Conference", *Chinese Academy of Sciences*, Beijing, China, Volumes. 1 and 2., 1995.

Naunidhi Kaur, "Privatising Water," *Frontline*, vol. 20, issue 18, Aug 30- Sept 12, 2003

Nityanand Jayaraman, "No water? Drink Coke", *CorpWatch India*, May 28, 2002,

Oweis, T. and Taimeh, A., *Overall Evaluation of On-Farm Water the Arid Regions*, In: Lacirignola, C. and A. Hamdy (eds) Proceedings, CIHEAM Conference "Land and Water Resources Management in the Mediterranean Region" 04. - 08, September, 1994

Oweis, T., A. Hachum, and J. Kijne, "Water harvesting and supplementary irrigation for improved water use efficiency in dry areas", *SWIM Paper 7*, Colombo, Sri Lanka: International Water Management Institute, 1999.

Pacey, A. & Cullis, A., *Rainwater Harvesting: The Collection of Rainfall and Runoff in Rural Areas*, London: I.T. Publications, 1986.

Pandey, Deep Narayan, Anil K. Gupta and David M. Anderson, "Rainwater harvesting as an adaptation to climate change", *Review article,* Current science, Vol. 85, No. 1, 10 July, 2003.

Papadopoulos, I., "Water Harvesting from Greenhouses", *In: FAO, Water Harvesting For Improved Agricultural Production*, Expert Consultation, Cairo, Egypt 21-25 Nov. 1993, p. 97-101, FAO, Rome.

Pearce, F., *The Dammed: Rivers, Dams and the coming world water crisis*, London: The Bodley Head, 1992.

Prinz, D., Tauer, W., Voegtle, "The Application of Geographic Information Systems to Identify Areas Suitable for Water Harvesting/Runoff Irrigation" *Proceedings, XII CIGR World Congress*, Milano 29.08 - 01.09.1994.

Roger P., "Integrated urban water resources management", *Natural Resources Forum* 17(1): 33-42, 1993

Sandweiss, D. H. *et. al.*, "Variation in Holocene El Nino frequencies: Climate records and cultural consequences in ancient Peru", *Geology*, 29, 603–606, 2001.

Scarborough, V. L. and Gallopin, G. G., "A water storage adaptation in Maya lowlands", *Science*, 251, 658–662, 1991.

Schreier, H. et. al., "Environmental Auditing—Gaining Forests But Losing Ground: A GIS Evaluation in a Himalayan Watershed", *Environmental Management* 18, no. 1,: 139–150, 1994.

Schultz, Gert A., "Use of Remote Sensing Data in a GIS Environment for Water Resources Management", *IAHS publication* 242: 3, 1997.

Scudder T., "Recent experiences with river basin development on the tropics", *Natural Resources Forum* 18(2): 101-114, 1994

Shankari Uma, Shah Esha, *Water Management Traditions in India*, Madras, PPST Foundations, 193 p., 1993.

Shatta, A. and F. Attia, "Environmental Aspects of Water Harvesting", *In: FAO, Water Harvesting For Improved Agricultural Production. Expert Consultation*, Cairo, Egypt 21-25 Nov. 1993.

Siegert, C., "Socio-economics of water harvesting", *Water Harvesting for Improved Agricultural Production.* Expert Consultation, Cairo, Egypt 21-25 Nov. 1993.

Siegert, K., "Introduction to Water Harvesting. Some Basic Principles for Planning", *Design and Monitoring. In: FAO, Water Harvesting For Improved Agricultural Production*, Expert Consultation, Cairo, Egypt 21-25 Nov. 1993.

Sonu Jain, "A River gone private is drying up" *The Sunday (Indian) Express*, April 13, 2003

Srinivasan, R., Arnold, J. G., "Integration of a Basin-Scale Water Quality Model With GIS", *Water Resources Bulletin* 30, no. 3: 453, 1994.

Sudhirendar Sharma, "Water Resources: Background and Perspective," InfoChange News and Features, April 2003

Tayaa, M., "Present Situation and Prospects for Improvement of Water Harvesting Practices in Morocco", *In: FAO, Water Harvesting For Improved Agricultural Production*, Expert Consultation, Cairo, Egypt 21-25 Nov. 1993.

Thomas, T., *Domestic Water Supply Using Rainwater Harvesting, Building Research and Information* Vol 2, No 2: 94-101, 1998.

Tobbi, B., "Water Harvesting: Historic, Existing and Potentials in Tunisia", *In: FAO, Water Harvesting For Improved Agricultural Production*, *Expert Consultation*, Cairo, Egypt 21-25 Nov. 1993.

Tsihrintzis, V. A., Hamid, R., Fuentes, H. R., "Use of Geographic Information Systems (GIS) in Water Resources: A Review", *Water Resources Management* 10, no. 4: 251, 1996.

UNEP, *Rain and Stormwater Harvesting in Rural Areas*, Tycooly, Dublin, 1983.

United Nations Environment Programme / Tycooly International Publishing Ltd., Rain and Stormwater Harvesting in Rural Areas, 238pp. *Water Resources Series*, 5. Nairobi and Dublin: United Nations Environment Programme / Tycooly International Publishing Ltd., 1983.

Van Dijk, J. and C. Reij, "Indigenous Water Harvesting Techniques in Sub-Saharan Africa: Examples from Sudan and the West African Sahel", *In: FAO, Water Harvesting For Improved Agricultural Production*, Expert Consultation, Cairo, Egypt 21-25 Nov. 1993.

Vandana Shiva, "Water Privatization in India," *Article in Public Citizen website*, July 31, 2002

Vijayalakshmi, K., K. P. R. Vittal and L. R. P. Singh, "Water Harvesting and Re-use". *In: ICAR. A Decade of Dryland Agricultural Research in India 1971-1980*, p. 103-119. New Delhi, India, 1982.

Webb. P. Iskandarani M., "Water Insecurity and the Poor", Issues and Research Needs", ZEF Bonn Center for Development Research University of Bonn, 1998.

Zohar, Y., J.A. Aronson and H.Loevenstein, "Fuelwood and Fodder Crops in Rainfall Harvesting Systems Under Arid Conditions" *A Research Progress Report*, ARO Volcani Center. Bet Dagan, Israel, 1987.